Research Progress in High-Performance Magnesium Alloy and Its Applications

Research Progress in High-Performance Magnesium Alloy and Its Applications

Editor

Di Wu

Basel • Beijing • Wuhan • Barcelona • Belgrade • Novi Sad • Cluj • Manchester

Editor
Di Wu
Institute of Metal Research,
Chinese Academy of Sciences
Shenyang, China

Editorial Office
MDPI
St. Alban-Anlage 66
4052 Basel, Switzerland

This is a reprint of articles from the Special Issue published online in the open access journal *Materials* (ISSN 1996-1944) (available at: https://www.mdpi.com/journal/materials/special_issues/RP_MA).

For citation purposes, cite each article independently as indicated on the article page online and as indicated below:

Lastname, A.A.; Lastname, B.B. Article Title. *Journal Name* **Year**, *Volume Number*, Page Range.

ISBN 978-3-0365-8594-9 (Hbk)
ISBN 978-3-0365-8595-6 (PDF)
doi.org/10.3390/books978-3-0365-8595-6

© 2023 by the authors. Articles in this book are Open Access and distributed under the Creative Commons Attribution (CC BY) license. The book as a whole is distributed by MDPI under the terms and conditions of the Creative Commons Attribution-NonCommercial-NoDerivs (CC BY-NC-ND) license.

Contents

About the Editor vii

Di Wu and Jinguo Li
Research Progress in High-Performance Magnesium Alloy and Its Applications
Reprinted from: *Materials* 2023, 16, 5460, doi:10.3390/ma16155460 1

Chengyu Zhang, Di Wu, Yanda He, Wenyu Pan, Jianqiu Wang and Enhou Han
Twinning Behavior, Microstructure Evolution and Mechanical Property of Random-Orientated ZK60 Mg Alloy Compressed at Room Temperature
Reprinted from: *Materials* 2023, 16, 1163, doi:10.3390/ma16031163 5

Songhe Lu, Di Wu, Ming Yan and Rongshi Chen
Achieving High-Strength and Toughness in a Mg-Gd-Y Alloy Using Multidirectional Impact Forging
Reprinted from: *Materials* 2022, 15, 1508, doi:10.3390/ma15041508 19

Yinyang Wang, Chen Liu, Yu Fu, Yongdong Xu, Zhiwen Shao, Xiaohu Chen and et al.
Simultaneously Improving Ductility and Stretch Formability of Mg-3Y Sheet via High Temperature Cross-Rolling and Subsequent Short-Term Annealing
Reprinted from: *Materials* 2022, 15, 4712, doi:10.3390/ma15134712 31

Xuefei Zhang, Baoyi Du and Yuejie Cao
Study on Microstructural Evolution and Mechanical Properties of Mg-3Sn-1Mn-xLa Alloy by Backward Extrusion
Reprinted from: *Materials* 2023, 16, 4588, doi:10.3390/ma16134588 61

Qinghang Wang, Li Wang, Haowei Zhai, Yang Chen and Shuai Chen
Establishment of Constitutive Model and Analysis of Dynamic Recrystallization Kinetics of Mg-Bi-Ca Alloy during Hot Deformation
Reprinted from: *Materials* 2022, 15, 7986, doi:10.3390/ma15227986 77

Vadym Shalomeev, Galyna Tabunshchyk, Viktor Greshta, Kinga Korniejenko, Martin Duarte Guigou and Sławomir Parzych
Casting Welding from Magnesium Alloy Using Filler Materials That Contain Scandium
Reprinted from: *Materials* 2022, 15, 4213, doi:10.3390/ma15124213 95

Junjian Fu, Wenbo Du, Ke Liu, Xian Du, Chenchen Zhao, Hongxing Liang and et al.
Effect of the $Ca_2Mg_6Zn_3$ Phase on the Corrosion Behavior of Biodegradable Mg-4.0Zn-0.2Mn-xCa Alloys in Hank's Solution
Reprinted from: *Materials* 2022, 15, 2079, doi:10.3390/ma15062079 107

Yusong Ma, Kaichuang Zhang, Shizhou Ma, Jinyan He, Xiqiang Gai and Xinggao Zhang
Ignition and Combustion Characteristic of B·Mg Alloy Powders
Reprinted from: *Materials* 2022, 15, 2717, doi:10.3390/ma15082717 121

Xuan Guo, Yunpeng Hu, Kezhen Yuan and Yang Qiao
Review of the Effect of Surface Coating Modification on Magnesium Alloy Biocompatibility
Reprinted from: *Materials* 2022, 15, 3291, doi:10.3390/ma15093291 131

Yan Liu, Zhaozhen Liu, Guishen Zhou, Chunlin He and Jun Zhang
Microstructures and Properties of Al-Mg Alloys Manufactured by WAAM-CMT
Reprinted from: *Materials* 2022, 15, 5460, doi:10.3390/ma15155460 149

About the Editor

Di Wu

   Di Wu is a professor at Institute of Metal Research, Chinese Academy of Sciences. He received his B.E. in 2006 and master's degree in 2008 from Northeastern University in China. Then, WU was awarded his Ph.D. degree in Materials Science and Engineering from the Institute of Metal Research (IMR), Chinese Academy of Sciences. After obtaining his Ph.D. degree, he joined IMR in 2012. Over the last 17 years, he has dedicated most of his efforts to the research and application of microstructure and properties regulation of Mg alloy and component forming process. To date, he has published over 70 peer-reviewed articles in journals including Acta Materialia; Scripta Materialia; Journal of Magnesium Alloys; and Journal of Materials Science and Technology.

Editorial

Research Progress in High-Performance Magnesium Alloy and Its Applications

Di Wu [1,2] and Jinguo Li [1,2,*]

[1] Shi-Changxu Innovation Center for Advanced Materials, Institute of Metal Research, Chinese Academy of Sciences, 72 Wenhua Road, Shenyang 110016, China; dwu@imr.ac.cn
[2] School of Materials Science and Engineering, University of Science and Technology of China, 72 Wenhua Road, Shenyang 110016, China
* Correspondence: jgli@imr.ac.cn

Magnesium is abundant in the Earth's crust and seawater. Mg alloy is the lightest metallic structural material, with the advantages of high specific strength, high specific stiffness, good electromagnetism shield, good damping capacity, good machinability, easy recycling, etc. Therefore, it has extremely broad application prospects and has drawn considerable interest in the automobile, electronics, electrical appliance, transportation, aerospace, aviation, and the national defense military industries. This Special Issue (SI), "Research Progress in High-Performance Magnesium Alloy and Its Applications", presents recent developments and excellent results in the field of Mg alloys, and includes ten articles covering some interesting and hot aspects of the topic. The purpose of the current Editorial is to briefly summarize the publications included in this SI.

Plastic processing is a promising method for improving the mechanical properties of metallic materials. Wrought Mg alloy usually exhibits superior strength and shows ample application potential. However, Mg alloy seems inferior in plastic deformation due to its HCP crystal structure, making it unable to provide sufficient independent slip systems. This SI fortunately covers the recent advances in the regulation of microstructure and mechanical properties during three traditional plastic processes, i.e., rolling, extruding, and forging. Twinning is the crystallographic shear process in grain which can change the grain orientation for a certain angle. Twin boundaries also can divide the matrix and generate grain refinement. At the same time, twin boundaries as two-dimensional lattice defects can be recrystallization nucleation sites, which facilitate recrystallization. It is evident that twinning can be used to improve the mechanical properties of Mg alloys. Zhang et al. [1] found that {10–12} twinning is one of the main deformation mechanisms of cast ZK60 Mg alloy during uniaxial compression at room temperature. It plays an important role in the evolution of microstructure, texture, and mechanical properties. Lu et al. [2] further developed a high-strength and -toughness Mg-Gd-Y alloy via MDIF (multidirectional impact forging) using {10–12} twin and correlated recrystallization. The forged sample had a fine-grained microstructure with an average grain size of ~5.7 μm and a weak nonbasal texture, and showed a high TYS (tensile yield strength) of 337 MPa, an EL (elongation) of 11.5%, and a ST (static toughness) of 50.4 MJ/m^3. It also presented yield isotropy (the ratio of compression yield strength/tensile yield strength along the forging direction was ≈1.0). Wang et al. [3] investigated the flow behavior of solution-treated Mg-3.2Bi-0.8Ca (BX31, wt.%) alloy during hot compression under different deformation conditions, and made hot processing maps for confirming a suitable hot working range. With the assistance of a hot processing map, the as-extruded alloy exhibited a smooth surface, a fine DRX structure with weak off-basal texture, and good strength–ductility synergy. In addition, Zhang et al. [4] prepared Mg-3Sn-1Mn-xLa alloy bars using backward extrusion, and systematically studied the effects of the La content on the microstructure and mechanical properties of the alloy. With the addition of La, the

Mg$_2$Sn phases exhibited significant refinement and spheroidization, and the grain size was significantly refined. Therefore, the mechanical properties of the extruded Mg-3Sn-1Mn-xLa alloy were significantly enhanced. A Mg alloy rolling sheet usually exhibits a strong basal texture and bad formability. Wang et al. [5] applied multipass high-temperature cross-rolling with interpass annealing to Mg-3Y alloy. The Mg-3Y alloy sheet presented a complete SRXed microstructure consisting of uniform equiaxed grains and a weakened multiple-peak texture. Systemic characterization and analysis indicated that the enhanced activity of basal <a> slip and randomized grain orientation played a significant role in decreasing the anisotropy of the Mg-3Y alloy sheet, which contributed to the formation of high stretch formability (~6.2 mm) at room temperature.

Mg alloys have attracted great attention as promising biodegradable materials for orthopedic implants and cardiovascular interventional devices, but the degradation rate is unbalanced due to their poor corrosion resistance in a physiological environment, which seriously affects their clinical use. Adding alloying elements and surface modification technology are two main ways to improve the corrosion resistance of Mg alloys. From the perspective of corrosion resistance and the biocompatibility of biomedical magnesium alloy materials, Guo et al. [6] reviewed the application and characteristics of six different surface-coating modifications in the biomedical magnesium alloy field, including the chemical conversion method, microarc oxidation method, sol–gel method, electrophoretic deposition, hydrothermal method, and thermal spraying method, and looked ahead towards the development prospects of surface-coating modification. Moreover, Fu et al. [7] investigated the effect of Ca addition on the corrosion behavior of biodegradable Mg-4.0Zn-0.2Mn alloys in Hank's solution. It was suggested that the Ca$_2$Mg$_6$Zn$_3$ acted as a cathode to accelerate the corrosion process due to the microgalvanic effect.

Although it is generally believed that Mg alloys have excellent casting properties, the complexity of aerospace structures and imperfect casting technology require the usage of additive welding technologies. Shalomeev et al. [8] developed a scandium-containing filler metal from a Mg-Zr-Nd system alloy for the welding of aircraft castings. The proposed filler material composition with an improved set of properties for the welding of body castings from a Mg-Zr-Nd system alloy for aircraft engines makes it possible to increase their reliability and durability in general, extend the service life of aircraft engines, and gain economic benefits. As a combustible metal powder, Mg may also be used for the development of advanced weapons and equipment. Ma et al. [9] fabricated MgB$_2$ via a combination of mechanical alloying and heat treatment, and found that its ignition temperature was greatly reduced in comparison with boron, which suggests that MgB$_2$ may be used in gunpowder, propellant, explosives, and pyrotechnics due to its improved ignition performance. Additive manufacturing (AM) is largely capable of manufacturing structures with high complexities. Liu et al. [10] manufactured Al-Mg alloy walls via WAAM (wire and arc additive manufacturing), which showed better performance than those produced using the traditional casting process under the optimal process parameters.

Author Contributions: Investigation, writing & editing, D.W.; supervision & review, J.L. All authors have read and agreed to the published version of the manuscript.

Funding: This work was funded by the National Natural Science Foundation of China (NSFC) through Projects No. 52171055 and No. 51301173.

Institutional Review Board Statement: Not applicable.

Informed Consent Statement: Not applicable.

Conflicts of Interest: The authors declare no conflict of interest.

References

1. Zhang, C.; Wu, D.; He, Y.; Pan, W.; Wang, J.; Han, E. Twinning Behavior, Microstructure Evolution and Mechanical Property of Random-Orientated ZK60 Mg Alloy Compressed at Room Temperature. *Materials* **2023**, *16*, 1163. [CrossRef] [PubMed]

2. Lu, S.; Wu, D.; Yan, M.; Chen, R. Achieving High-Strength and Toughness in a Mg-Gd-Y Alloy Using Multidirectional Impact Forging. *Materials* **2022**, *15*, 1508. [CrossRef] [PubMed]
3. Wang, Q.; Wang, L.; Zhai, H.; Chen, Y.; Chen, S. Establishment of Constitutive Model and Analysis of Dynamic Recrystallization Kinetics of Mg-Bi-Ca Alloy during Hot Deformation. *Materials* **2022**, *15*, 7986. [CrossRef] [PubMed]
4. Zhang, X.; Du, B.; Cao, Y. Study on Microstructural Evolution and Mechanical Properties of Mg-3Sn-1Mn-xLa Alloy by Backward Extrusion. *Materials* **2023**, *16*, 4588. [CrossRef] [PubMed]
5. Wang, Y.; Liu, C.; Fu, Y.; Xu, Y.; Shao, Z.; Chen, X.; Zhu, X. Simultaneously Improving Ductility and Stretch Formability of Mg-3Y Sheet via High Temperature Cross-Rolling and Subsequent Short-Term Annealing. *Materials* **2022**, *15*, 4712. [CrossRef] [PubMed]
6. Guo, X.; Hu, Y.; Yuan, K.; Qiao, Y. Review of the Effect of Surface Coating Modification on Magnesium Alloy Biocompatibility. *Materials* **2022**, *15*, 3291. [CrossRef] [PubMed]
7. Fu, J.; Du, W.; Liu, K.; Du, X.; Zhao, C.; Liang, H.; Mansoor, A.; Li, S.; Wang, Z. Effect of the Ca2Mg6Zn3 Phase on the Corrosion Behavior of Biodegradable Mg-4.0Zn-0.2Mn-xCa Alloys in Hank's Solution. *Materials* **2022**, *15*, 2079. [CrossRef] [PubMed]
8. Shalomeev, V.; Tabunshchyk, G.; Greshta, V.; Korniejenko, K.; Guigou, M.D.; Parzych, S. Casting Welding from Magnesium Alloy Using Filler Materials That Contain Scandium. *Materials* **2022**, *15*, 4213. [CrossRef]
9. Ma, Y.; Zhang, K.; Ma, S.; He, J.; Gai, X.; Zhang, X. Ignition and Combustion Characteristic of B·Mg Alloy Powders. *Materials* **2022**, *15*, 2717. [CrossRef] [PubMed]
10. Liu, Y.; Liu, Z.; Zhou, G.; He, C.; Zhang, J. Microstructures and Properties of Al-Mg Alloys Manufactured by WAAM-CMT. *Materials* **2022**, *15*, 5460. [CrossRef] [PubMed]

Disclaimer/Publisher's Note: The statements, opinions and data contained in all publications are solely those of the individual author(s) and contributor(s) and not of MDPI and/or the editor(s). MDPI and/or the editor(s) disclaim responsibility for any injury to people or property resulting from any ideas, methods, instructions or products referred to in the content.

Article

Twinning Behavior, Microstructure Evolution and Mechanical Property of Random-Orientated ZK60 Mg Alloy Compressed at Room Temperature

Chengyu Zhang [1,2], Di Wu [1,2,*], Yanda He [1,3], Wenyu Pan [1,4], Jianqiu Wang [1,5,*] and Enhou Han [1,5]

1. CAS Key Laboratory of Nuclear Materials and Safety Assessment, Institute of Metal Research, Chinese Academy of Sciences, 62 Wencui Road, Shenyang 110016, China
2. School of Materials Science and Engineering, University of Science and Technology of China, 72 Wenhua Road, Shenyang 110016, China
3. School of Nano Science and Technology, University of Science and Technology of China, 166 Renai Road, Suzhou 215127, China
4. School of Materials Science and Engineering, Shenyang University of Chemical Technology, 11th Street, Shenyang Economic and Technological Development, Shenyang 110142, China
5. Institute of Corrosion Science and Technology, 136 Kaiyuan Avenue, Guangzhou 510530, China
* Correspondence: dwu@imr.ac.cn (D.W.); wangjianqiu@imr.ac.cn (J.W.)

Abstract: In this study, the uniaxial compression of random orientation ZK60 Mg alloy to different strains was performed at room temperature. The microstructure evolution was characterized mainly using electron backscattered diffraction (EBSD), and the mechanical property was evaluated by the Vickers hardness test. During compression, extension twins nucleated, grew, and engulfed the grain. Twins form a texture with the c-axis parallel to the compression direction. With the massive nucleation and expansion of extension twins during compression, the twin boundary (TB) brought the grain refinement, and the twin boundary-dislocation interaction significantly increased the strain hardening rate of ZK60 Mg alloy, both leading to its significantly increasement of the hardness.

Keywords: Mg alloy; twinning; random orientation; compression; room temperature

1. Introduction

Mg alloy is the lightest structural metal, which can play an important role in energy saving and emission reduction. In industry, over 80% of Mg components are made by high-pressure die casting [1], but casting brings defects such as porosity and shrinkage cavity [2,3]; these defects reduce the consistency of mechanical properties and limit the widespread use of Mg alloys. In contrast, the Mg alloy processed by rolling, extrusion, and forging can reduce casting defects, refine the structure and improve mechanical properties [4–7]. Some wrought Mg alloys made by rotary swaging have a strength of over 700 MPa [5,8], which shows the ample potential of Mg alloy for application.

As a result, studying the microstructure evolution during the plastic forming is critical. In the mechanical deformation process, dislocation slipping and twinning are the main deformation methods. For dislocation slipping, non-basal slip is difficult to activate at room temperature. This phenomenon makes the Mg alloy prone to cracking during cold working [9]. Then, thermal processing of magnesium alloys was developed to promote the activation of non-basal slip by increasing the processing temperature. Although the activation of non-basal slip can be promoted by heating, Mg alloy has a high thermal conductivity, which is significantly higher than aluminum alloy and steel [10–12]. This physical property causes the temperature of Mg alloy dropping rapidly during processing, which results in a narrow processing window for Mg alloy, making it difficult to ensure that the Mg alloy ingots are deformed within the ideal temperature range. When the temperature is below the processing window, the large strain will lead to cracking and strength inhomogeneity of the material [13]. In contrast, the CRSS of twinning is almost independent of temperature, so the control of Mg alloy microstructure by twinning at room temperature becomes the preferred choice.

Twinning is the crystallographic shear process in grain which can change the grain orientation for a certain angle. Twin boundary also can divide the matrix and brings grain refinement. At the same time, twin boundaries as two-dimensional lattice defects can be recrystallization nucleation sites, which facilitate recrystallization [14]. It is evident that twinning can be used to improve the mechanical properties of Mg alloys. As mentioned above, twinning can effectively regulate the microstructure and improves the mechanical properties of Mg alloys. We can promote twinning by reducing the temperature since the CRSS of twinning is less affected by temperature. The CRSS of {10-12} extension twinning is close to the basal slip in Mg alloy [15], which means the {10-12} twinning can be activated easily. This means that it is feasible to regulate the microstructure of Mg alloys by twinning.

In previous studies, it has been found that there are many factors affecting the twinning behavior of Mg alloys, such as temperature [16], alloy composition [17–19], strain [20], etc. Therefore, we need a lot of systematic research to regulate the twinning behavior in Mg alloys [21]. Although some studies have been carried out on twinning in Mg alloys, the twinning behavior is not fully investigated at present, and twinning behavior is complex and needs to be studied in more depth. For example, rare earth elements can significantly inhibit twinning and producing the unusual {11-21} twins [22,23]. In addition, the adding of reinforcing phase in the magnesium alloy can effectively promote the nucleation of twin variants [24].

Strain also significantly affects the evolution of twinning behavior of Mg alloys during processing, which will have a significant effect on the final properties of Mg alloys. Many twinning behavior of single-component Mg alloys has been studied, but it is not detailed and thorough enough [25,26]. In order to make full use of the regulation of Mg alloy twinning on the microstructure, more studies are needed on the twinning behavior of Mg alloys from small strains to failure. In addition, the current studies are based on hot-worked, strong-textured Mg alloys, and there is a lack of a specific study of a particular Mg alloy, especially in different strains.

ZK60 Mg alloy is the typical high-strength wrought Mg alloy, and advanced processing has been widely used on it [27,28]. In this article, the cast ZK60 Mg alloy with random orientation was selected, and deformed at room temperature. EBSD was mainly used to analyze its twinning behavior, and related microstructure evolution, texture and mechanical properties were also investigated.

2. Materials and Methods

2.1. Materials Preparation

The material used in this research is a commercial cast ZK60 (Mg-Zn-Zr) alloy in homogenized state; Table 1 presents the chemical composition of this alloy. In this study, the cylinder compression sample was used, with dimensions $\Phi 10 \times 15$ mm, machined by wire electrical discharge machining.

Table 1. The chemical composition of ZK60 Mg alloy.

	Mg	Zn	Zr	Y
wt%	Bal.	5.56	0.58	<0.01

2.2. Microstructure Characterization

The phase composition of the specimen was identified by X-ray diffraction analysis (XRD) (Rigaku D/Max-2500 PC, Tokyo, Japan) with a scan speed of 2°/min. The metallography was observed by optical microscope (OM) (Zeiss Axio Observer Z1, Jena, Germany). The microstructure and the element distribution were observed and analyzed by environmental scanning electron microscope (ESEM) (FEI XL30 FEG, Eindhoven, The Netherland) equipped with energy dispersive spectrometer (EDS) (EDAX, Oxfordshire, UK). Transmission electron microscope (TEM) (JEOL J2100F, Tokyo, Japan) was also used to analyze the detailed microstructure of the ZK60 Mg alloy.

2.3. Electron Backscatter Diffraction Analyzation

The electron backscatter diffraction (EBSD) specimens were firstly grinded with abrasive paper from 200# to 5000#, and electrochemical polishing was used at 15 V with the electrolyte (90% ethanol and 10% perchloric acid) for the 30 s. The EBSD analysis was conducted by ESEM (Thermo Scientific Quattro S, Waltham, MA, USA) equipped with EBSD detector (EDAX, DigiView, Mahwah, MA, USA). A 20 KeV electron beam was selected with a spot size of 6.5, and the step size was 2.5 μm. The EBSD data was analyzed by TSL OIM software (Version 8.62).

2.4. Room Temperature Compression

The specimens were compressed at room temperature with a strain rate of 0.001 s^{-1}. The compression experiment works on the universal testing machine (Shimadzu, AG-X 50 kN, Kyoto, Japan). A schematic of the experimental approach and sampling observation is illustrated in Figure 1, and the reference directions are defined as CD (compression direction) and TD (transverse detraction).

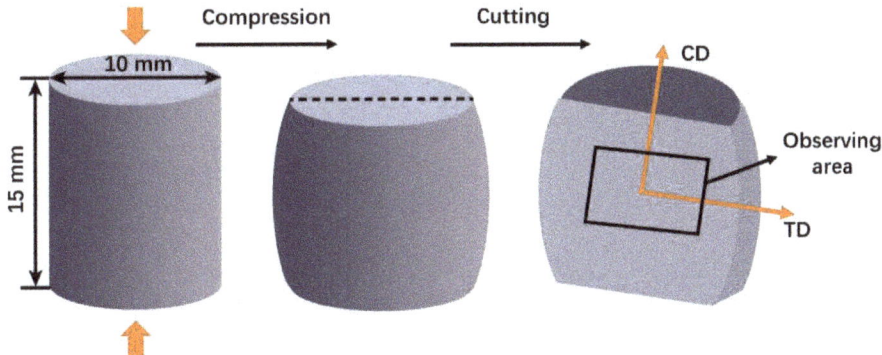

Figure 1. The compression scheme and EBSD sample coordinate system.

2.5. Hardness Test

The Vickers hardness test used the MHVD-1000 AP microhardness tester (Shanghai Optics And Dine Mechanics Institution, Chinese Academy of Sciences, Shanghai, China) with an applied load of 500 g and a holding time of 15 s. For experimental confidence, every specimen was tested eight times at the core area.

3. Results

3.1. Microstructure Characterization of Initial ZK60 Alloy

Figure 2 shows the XRD pattern of the initial ZK60 alloy. The cast ZK60 alloy is mainly composed α-Mg, and a small amount of MgZn$_2$ and Zn$_2$Zr phase. Figure 3 shows the OM and SEM images of the as-homogenized alloy. The initial microstructure consists polygonal grains, and most grain size was over 200 μm. The petaloid patches were observed in the interior of the matrix, and some bulk-shaped phases were distributed randomly at the grain boundary. Figure 3b is the high magnification image of petaloid patches, and a high-density rod precipitated phase was observed.

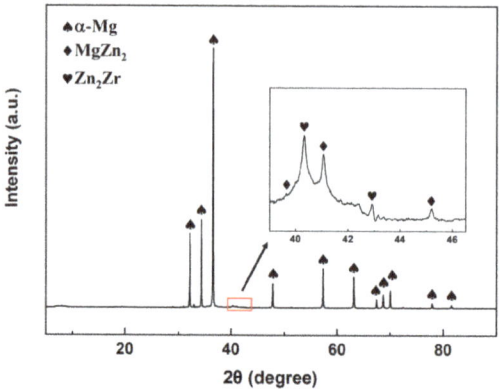

Figure 2. XRD pattern of cast ZK60 Mg alloy.

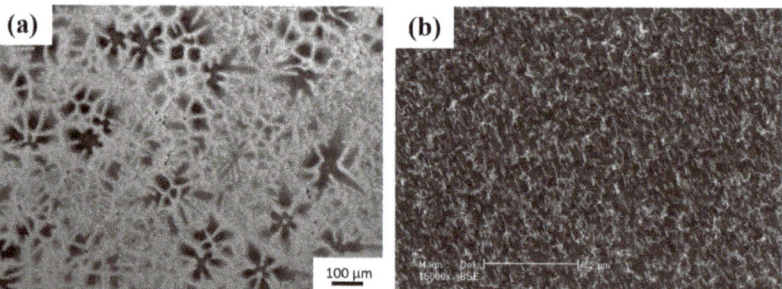

Figure 3. (**a**) Metallograph and (**b**) SEM image of cast ZK60 Mg alloy.

Figure 4a is the TEM image of the initial ZK60 Mg alloy. The rod precipitated phase has the preferred orientation, which has been confirmed is parallel with the c-axis of the grain [29,30]. Figure 4b shows the bright filed STEM image and EDS mapping of dense distribution of rod phase. From the EDS mapping, we can see the Zn element congregate at the rod phase. Combined with XRD results, the rod-like precipitated phase can be identified as $MgZn_2$. Figure 5 shows the SEM-EDS results, and bulk-shaped phases at the grain boundary were Zn_2Zr and ZnZr, which was also identified in the XRD pattern.

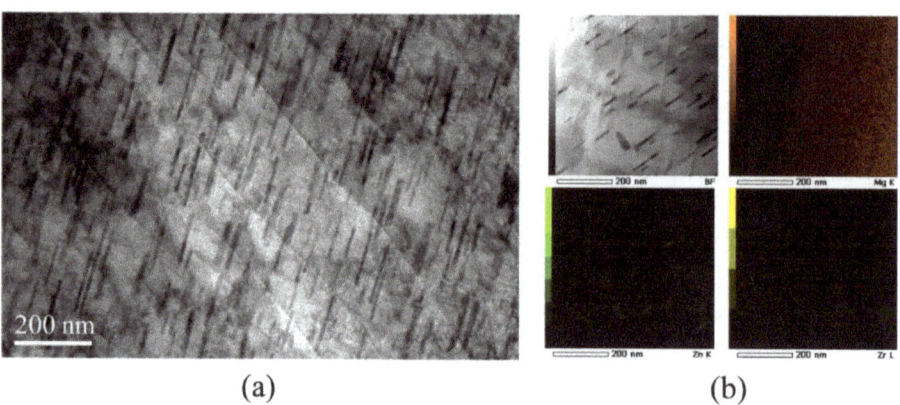

Figure 4. (**a**) TEM image and (**b**) STEM/EDS mapping of cast ZK60 Mg alloy.

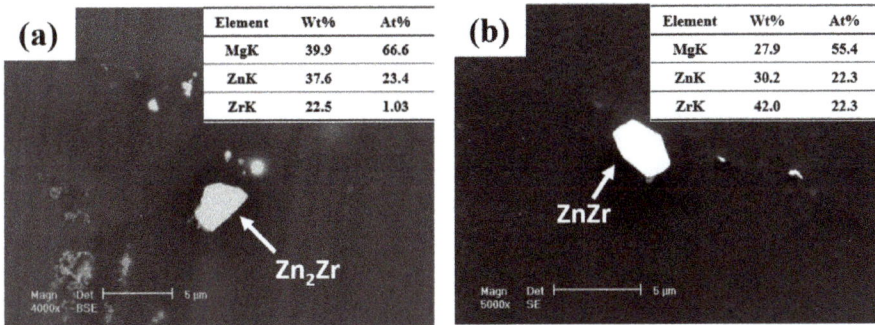

Figure 5. (a,b) the SEM image of particle phase at the grain boundary, and corresponding spot spectrum EDS results of bulk-shaped phases at grain boundary.

3.2. The Compression Curves

Figure 6 shows the true stress-strain curve and strain hardening rate-true strain curve. The concave down of the true stress-strain curve and the uplift of strain hardening rate-true strain curve all manifest the main deformation mechanism of ZK60 Mg alloy is {10-12} extension twinning [20,31,32]. Table 2 shows the strain of each point in Figure 6. For points P1 and P2, which are points with small strains, they are mainly used to observe the strain range in which twinning occurs and whether twinning occurs at small strains. For point P3, this is the point where the strain hardening rate appears to rise significantly near the middle of the range taken. Point P4 is located in the area where the strain hardening rate falls rapidly again after the highest point, and point P5 in the area where the decreasing trend of strain hardening rate slows down. For the region where P6 is located, there is a significant decrease in strain hardening rate and fracture may have occurred in this region.

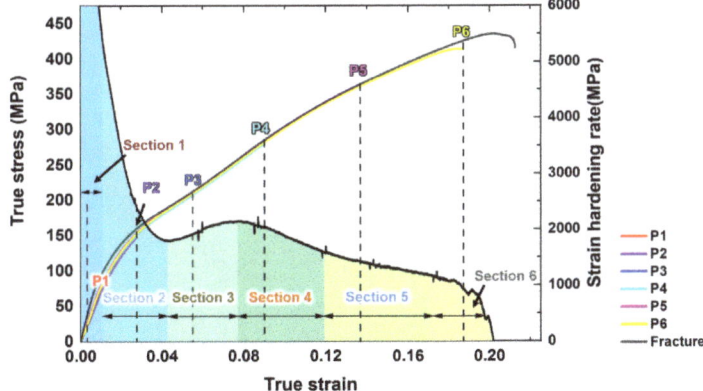

Figure 6. The true stress-strain curve and strain hardening rate-true strain curve.

Table 2. The six compression strain points for further microstructure characterization.

	P1	P2	P3	P4	P5	P6
True Strain	0.0025	0.0275	0.055	0.091	0.136	0.187
Engineering Strain	0.0025	0.0279	0.057	0.094	0.146	0.206

3.3. Microstructure Evolution

From Figure 7, the EBSD results show that the initial alloy has equiaxed grains with curved boundaries, and the pole figure (PF) shows the initial alloy exhibits low texture intensity and multi-peaks, which means the grains have a random orientation. Figure 7c shows that the grain size is mainly distributed from 140 to 240 μm, and the average grain size is 173.6 μm. Because of the noise and some tiny grains. The statistical grain size can have a small gap from the actual grain size. Figure 7d exhibits that the grain boundaries are mainly high-angle grain boundaries (HAGBs).

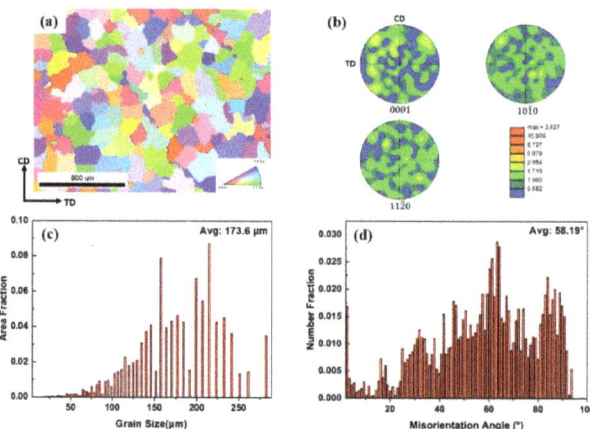

Figure 7. (**a**) The inverse pole figure (IPF) map, (**b**) pole figure, (**c**) grain size, and (**d**) misorientation angle distribution of as-homogenized ZK60 Mg alloy.

From Figure 8, the IPF map shows no twins nucleated at P1, but in metallography, small twins can be observed. The tiny morphology and the 2.5 μm step size keep the twins from being indexed by EBSD. At P1 the stress and train were 23.8 MPa and 0.25%. In Koike's work the {10-12} extension twin CRSS of polycrystalline Mg alloys is activated at 2–2.8 MPa [33]. Lu et al. found twin nucleates under the strain of 0.3% [34], and Chen et al. found the twins nucleated at the strain of 0.25% [35]. These previous works confirm that twins can be nucleated at low strain and stress.

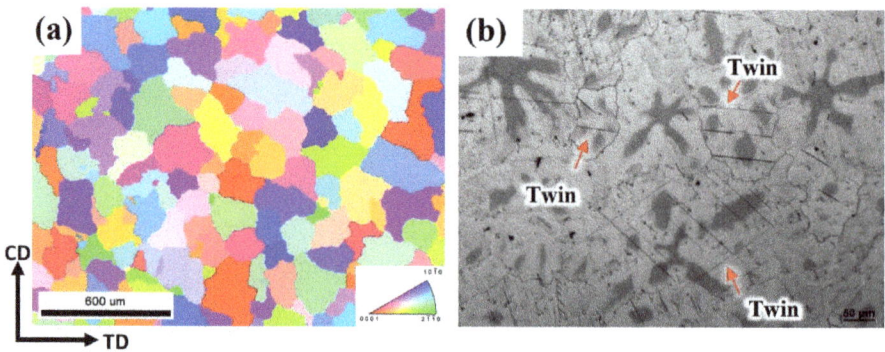

Figure 8. (**a**) The IPF map and (**b**) metallography at P1.

Figure 9 shows the IPF + IQ (image quality) + GB (Grain Boundary) map in the compression process. At P2, twinning can be detected by EBSD. Due to the low CRSS of {10-12} extension twinning, even if the grain orientation is not beneficial to the {10-12} extension twinning, they can still nucleate at P2 [36], and Figure 9a confirms that all twins at P2 are {10-12} extension twins. In P2's IPF map, we can see that the twins have parallel and cross structure. The cross structure is one kind of work hardening method of Mg alloy [37,38]. Moreover, such cross structure divides the matrix and produces a closed space restricting the dislocation motion [38] and the further expansion of twins [39]. These factors make the cross structure need more force to continue the deformation. At P3, many twins were expanding. In some areas, the twin can merge half the area of the matrix. At P4, some twins expanded to swallow the entire matrix, which means the matrix was covered by the twins, and most of the twin boundaries detected were still {10-12} extension twin boundary. Moreover, some thin twins can be found, and the morphology is the same as {10-11} compression twins. At P5, due to the large strains and stresses, a large number of dislocations interactions formed deformation band in the ZK60 Mg alloy. At P6, the deformation in the specimen is severe, and the index of EBSD is much lower than P1-P5 specimen, and more deformation bands can be observed in Figure 10b.

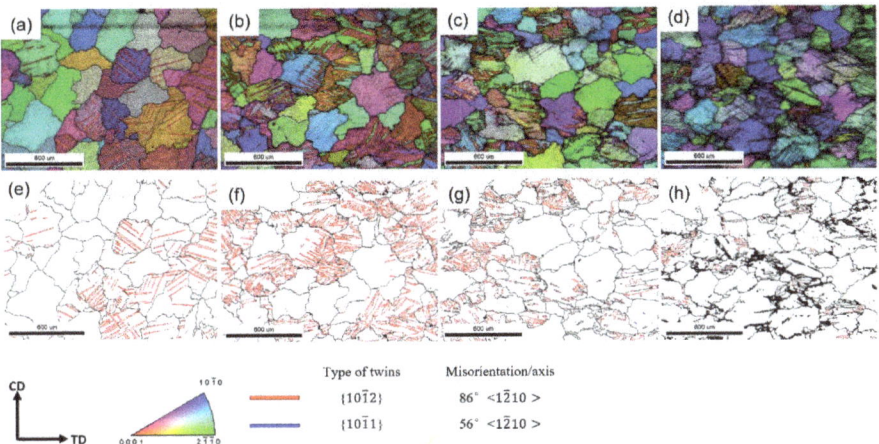

Figure 9. The IPF + IQ + GB map of the specimens at (**a**) P2, (**b**) P3, (**c**) P4, (**d**) P5, and the corresponding GB map of specimens at (**e**) P2, (**f**) P3, (**g**) P4, (**h**) P5.

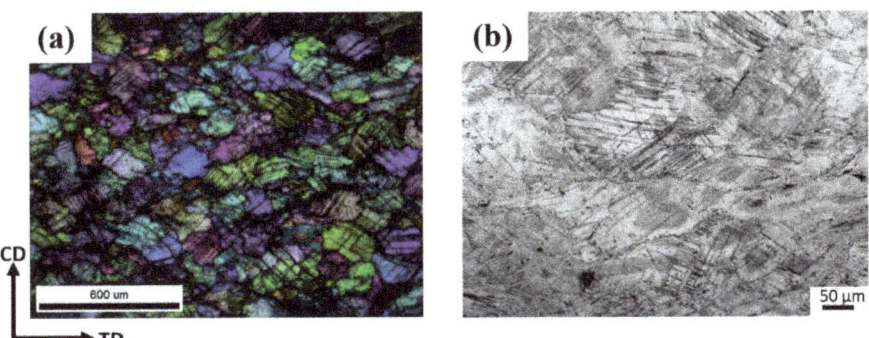

Figure 10. (**a**) The IPF map and (**b**) metallography at P6.

3.4. Orientation Statistics

From Figure 11, P1 shows a random distribution of misorientation angle, which means the microstructure was not explicit changed. At P2, the 86.5° misorientation angle increases rapidly, which means the nucleation and expansion of {10-12} extension twinning. At P3, the fraction of 86.5° decreased slightly, and the low-angle grain boundaries (LAGBs) increased. Compared with Figure 9, the expansion and merging of {10-12} extension twinning was the reason for the decrease of 86.5° peak. At P4, the peak of 86.5° decreased obviously, which means many grains were merged by the twins. Meanwhile, 56.2° misorientation appeared and it means the {10-11} compression twinning nucleate, which is a coincidence with Figure 9g. The LAGBs also increased, obviously. As the strain increased to P5, the misorientation of 86.5° almost disappeared, which means the twinned matrix was all merged by the {10-12} twins. In comparison with that at P5, the fraction of LAGBs at P6 changes less, and the peak of 56.2° becomes less obvious due to the increase in background intensity, and from Figure 10b, we can still see the thin twins.

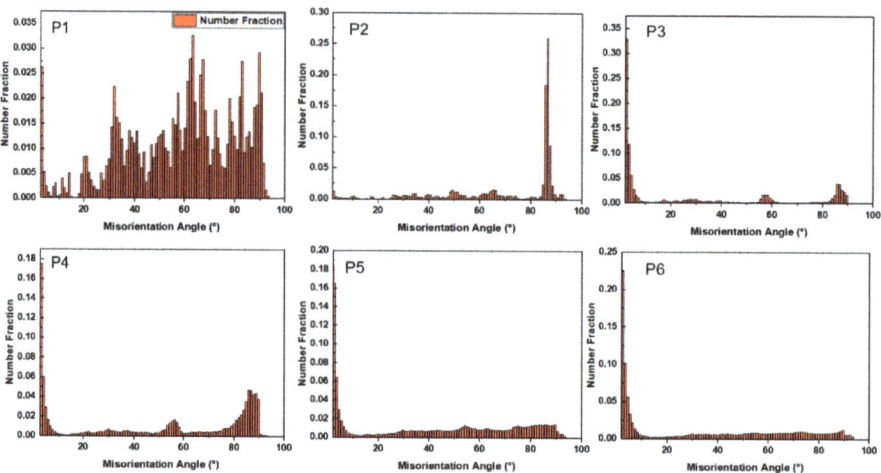

Figure 11. The misorientation angle at P1, P2, P3, P4, P5, and P6.

Table 3 shows that the LAGBs significantly increased than initial. From P1 to P3, the LAGBs increase gradually. From P5 to P6, the LAGBs slightly increased. In the compression process, the density of the dislocation increases with the increase in the strain, which means the entanglement and interaction of the dislocations also increased and formed the LAGBs.

Table 3. The fraction of LAGBs and 86.5° twin boundaries.

	P1	P2	P3	P4	P5	P6
LAGBs	0.053	0.032	0.569	0.308	0.332	0.479
86.5° ± 5°	0.170	0.576	0.676	0.414	0.181	0.154

Figure 12 shows the PFs of the ZK60 Mg alloy in the compression process. From the PFs, we can see the basal texture intensity of the specimens increased with the strain. At P4, the basal texture was stable, and P5 had the same texture distribution but higher intensity. The basal texture is the typical texture of Mg alloy where the c-axis of the grain is parallel to the compression direction. In Figure 13, we separated the twin and the matrix at the P3 strain, and it is clear that twinning is the direct reason for the basal texture in the compressed ZK60 Mg alloy, while the matrix maintains a relatively random orientation.

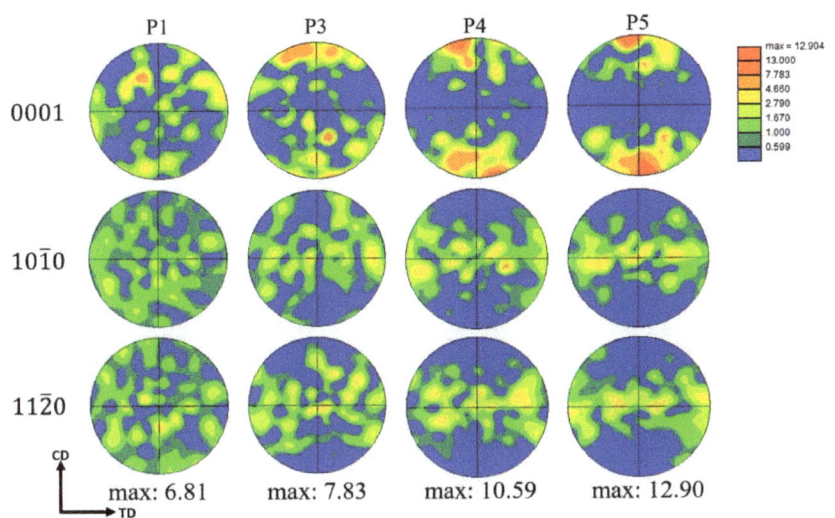

Figure 12. The PF at P1, P3, P4, and P5.

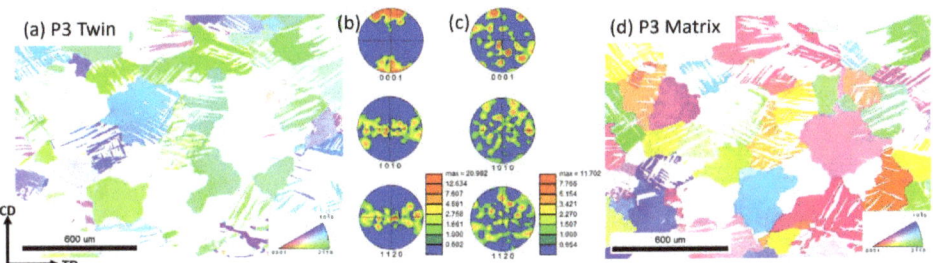

Figure 13. (**a**,**d**) The IPF Map and (**b**,**c**) corresponding PF of the twin and the matrix of P3.

3.5. The Evolution of the Schmid Factor

Figure 14 shows the Schmid Factor (SF) at P2 and P5. The basal slip SF has decreased slightly. Although the texture intensity of the Mg alloy increases significantly, the Schmid factor of base slip still has a high level, which means the basal slip remained sufficient to activate. After being compressed to P5, the SF of prismatic slip decreased a lot. Most of the grains have a low SF (<0.2) of prismatic slip, indicating the suppression to prismatic slip. Figure 14c shows the SF of pyramidal <c + a> slip becomes higher, which is similar to Gui et al.'s work [40], and the activation of pyramidal <c + a> was reasonable with the high stress at P5.

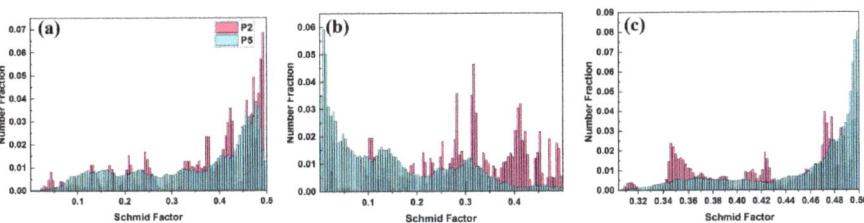

Figure 14. The Schmid factor at P2 and P5: (**a**) the basal slip, (**b**) the prismatic slip, (**c**) pyramidal <c + a> slip.

3.6. The Interaction between Twin and Participate Phase

From Figure 9, we can see that twins did not occur in some grains, the SEM was used to observe the grains. In Figure 15a, the thin twins nucleate with high density was shown. In order to investigate the reason of this phenomenon, TEM was used to observe the relationship between twinning and precipitation phases. From Figure 15b, the TEM image shows that the rod participates was not sheared by twin; it was kinked by a small angle which is the same as Robson's work [41], and much research shows that the participated phase will inhibit the expansion of twins, which will increase the stress in the matrix. The increase in stress will provide the additional drive force for twinning nucleation [30,41,42], which lead to the thin twins nucleate with high density shown in Figure 15a.

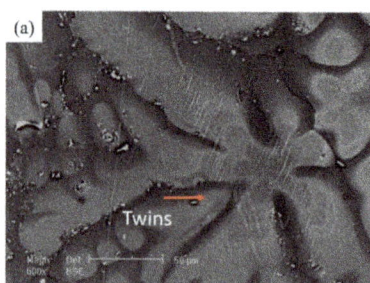 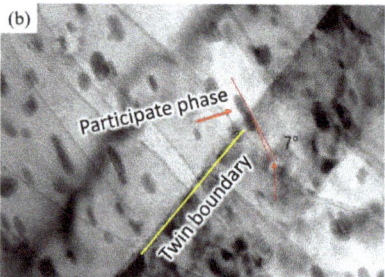

Figure 15. (a) The participate phase in the petaloid area, and (b) The TEM image of the participated phase at TB.

4. Discussion

4.1. The Relationship between Twinning and Strain Hardening at Room Temperature

From Section 3, we can divide the evolution of microstructure into four parts. (a) The nucleate of the {10-12} extension twins (P1–P2), (b) the growth of {10-12} extension twins (P2–P4), and (c) the deformation band formed (P4–P5). (d) Fracture (P6).

In the first part (P1–P2), there is an elasto-plastic transition at this stage and the slip is massively activated makes the decrease in strain hardening rate slow down [43–45]. In the second part (P2–P4), at P2, the twins grew up in many grains, as shown in Figure 9. The twins divide the matrix into many small pieces, which brings the grain refinement effect with obvious strain hardening. Figure 6 shows that, with the increase in the strain, some grains were merged by the twinning.

In Figure 16, Grain G3 has two different varieties of cross structure, which can lead to obvious strain hardening. Moreover, the different twin varieties cannot merge with each other, and their interaction will lead to more strain hardening [46]. In some grains, the twins grow fast, which have only one variety that will cover the matrix, as shown in Grain G1 in Figure 16. For Grain G1, we can clearly see the 86.5° {10-12} <11–20> TB between the matrix and the twin. Given the stress loading direction and the orientation of the matrix, we can see that the green area is the rest of the matrix, and the blue area is the twin, which means that the twinning is covering almost the whole grain, and the Grain G2 shows the same status.

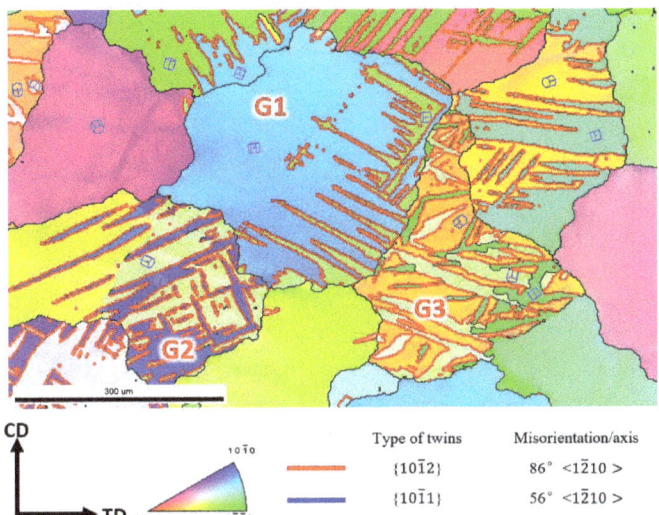

Figure 16. The IPF + GB map of the P3 specimen. G1 is the grain the matrix was almost covered by the twin in blue. G2 is the grain which matrix is yellow-green and twinning is blue. G3 is the matrix in yellow with two twin variants.

In the third part, the generation of extension twins leads to a significant basal texture in the ZK60 Mg alloy, which promote the activation of pyramid slip. The massive activation of pyramid slip provided enough strain hardening and make the strain hardening rate keep a stable value, and the accumulation of slip leads to the creation of deformation bands which was shown in Figure 9d.

In the fourth part, at P6, crakes occurred. In Figure 17, we can observe the crakes, and the contraction TBs usually were considered as the cracker source [46].

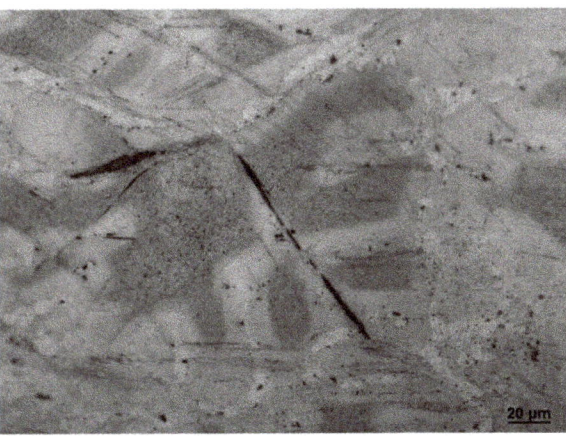

Figure 17. The crack in the P6 specimen.

4.2. The Mechanical Property Related to Twinning

Figure 18 shows a significant correlation between strain hardening rate and hardness. With the strain hardening rate decreasing rapidly, P1 shows that the hardness increased, then the hardness dropped a little; after P2, with the increase in the strain, the hardness continued to increase. In short, after compression, the hardness of the specimen continuously increases in comparison with the initial state.

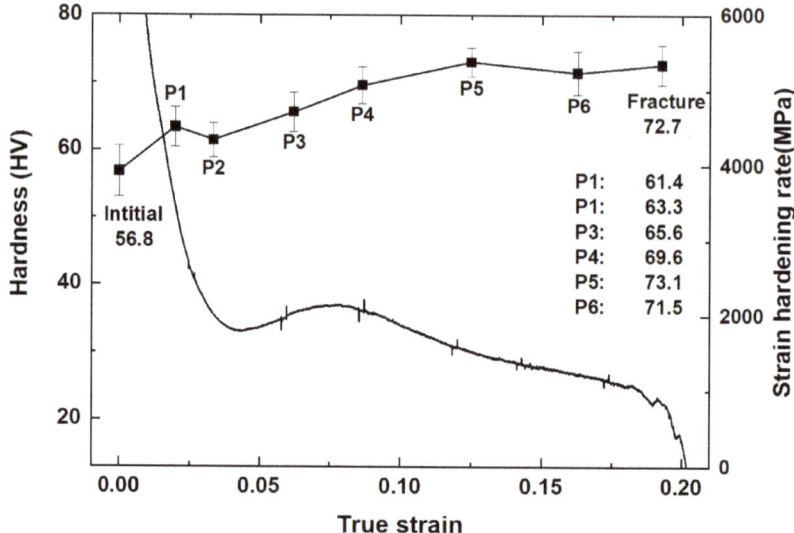

Figure 18. The hardness and strain hardening rate evolution with true strain.

The hardness evolution shows a strong relationship with the strain hardening rate. From Figures 6 and 18, we can clearly see that the in the area around P1, the hardness of the specimen increased rapidly, which means that the dislocation activation and the interaction between twinning and dislocation bring the obvious strain hardening. In this part, the stress was quite low (~50 MPa), so the slip activated should be the basal slip.

Then, from P2–P4, the massive nucleation of the twins and the length of TB increased rapidly, dividing the matrix and the TB-dislocation interaction, and the twin-twin interaction caused the hardness to increase rapidly. In P4–P5, the twins expand to the whole matrix, which means TB as the barrier potential of dislocation activation is lost, but in this section, the activation of the basal slip and pyramid slip makes the dislocations have a strong interaction, and also makes the alloy still have the high hardness.

5. Conclusions

Twinning is one of the main deformation mechanisms of cast ZK60 Mg alloy during uniaxial compression at room temperature. It plays an important role in the evolution of microstructure, texture, and mechanical property. Some detailed conclusions can be draw as follows:

(1) At the beginning of deformation, the {10-12} extension twinning is one of the main deformation mechanisms and is responsible for the basal texture. Accompanied by the formation of strong basal texture in the middle and late stage of deformation, compression twinning and deformation band occurred, which is the main source of crack initiation leading to failure.

(2) Slip is also an indispensable deformation mechanism. Basal slip still kept sufficient to activate during the whole compression process. With the strain increase and texture evolution, non-basal slip gradually turned from prismatic slip to pyramidal <c + a> slip.

(3) During compression, the interaction between the twins, and the interaction between twin boundaries and dislocations can significantly enhance the strain hardening rate of Mg alloy, and these two interactions together with the segmentation of grains by twin boundaries improve the mechanical properties of ZK60 alloy.

(4) The fine and dense precipitates of Mg-Zn phase in ZK60 Mg alloy will not be cut when sheared by the twin boundary, but is rotated by a small angle. The precipitation may hinder the growth of twins, but promote the nucleation of twinning.

Author Contributions: Investigation, writing—original draft preparation, C.Z.; funding acquisition, review and editing, D.W.; data curation, review and editing Y.H. and W.P.; supervision J.W. and E.H. All authors have read and agreed to the published version of the manuscript.

Funding: This research was funded by National Natural Science Foundation of China (NSFC) through Projects No. 52171055 and No. 51301173.

Institutional Review Board Statement: Not applicable.

Informed Consent Statement: Not applicable.

Data Availability Statement: The data presented in this study are available upon request from the corresponding author. The data are not publicly available due to the requirements of related projects.

Conflicts of Interest: The authors declare no conflict of interest.

References

1. Joost, W.J.; Krajewski, P.E. Towards magnesium alloys for high-volume automotive applications. *Scr. Mater.* **2017**, *128*, 107–112. [CrossRef]
2. Weiler, J.P.; Wood, J.T. Modeling the tensile failure of cast magnesium alloys. *J. Alloys Compd.* **2012**, *537*, 133–140. [CrossRef]
3. Gao, M.; He, G.H.; Huang, T.; Wang, C.Y.; Wu, C.; Yu, G.Y. Reducing the Shrinkage Defects in the ZA27 Casting Alloy by Using Electric Pulse Current. In Proceedings of the 3rd International Conference on Manufacturing Science and Engineering (ICMSE 2012), Xiamen, China, 27–29 March 2012; pp. 1082–1086.
4. Jiang, M.G.; Yan, H.; Gao, L.; Chen, R.S. Microstructural evolution of Mg-7Al-2Sn Mg alloy during multi-directional impact forging. *J. Magnes. Alloys* **2015**, *3*, 180–187. [CrossRef]
5. Yamashita, A.; Horita, Z.; Langdon, T.G. Improving the mechanical properties of magnesium and a magnesium alloy through severe plastic deformation. *Mater. Sci. Eng. A* **2001**, *300*, 142–147. [CrossRef]
6. Hong, M.; Wu, D.; Chen, R.S.; Du, X.H. Ductility enhancement of EW75 alloy by multi-directional forging. *J. Magnes. Alloys* **2014**, *2*, 317–324. [CrossRef]
7. Wang, T.; Zha, M.; Du, C.; Jia, H.-L.; Wang, C.; Guan, K.; Gao, Y.; Wang, H.-Y. High strength and high ductility achieved in a heterogeneous lamella-structured magnesium alloy. *Mater. Res. Lett.* **2023**, *11*, 187–195. [CrossRef]
8. Wan, Y.; Tang, B.; Gao, Y.; Tang, L.; Sha, G.; Zhang, B.; Liang, N.; Liu, C.; Jiang, S.; Chen, Z.; et al. Bulk nanocrystalline high-strength magnesium alloys prepared via rotary swaging. *Acta Mater.* **2020**, *200*, 274–286. [CrossRef]
9. Hwang, D.Y.; Shimamoto, A.; Kubota, R. A Study on the Fracture Behavior of Magnesium and Aluminium Alloy under Dynamic Biaxial Stress. *Key Eng. Mater.* **2005**, *297–300*, 1579–1584. [CrossRef]
10. Lee, S.; Ham, H.J.; Kwon, S.Y.; Kim, S.W.; Suh, C.M. Thermal Conductivity of Magnesium Alloys in the Temperature Range from −125 °C to 400 °C. *Int. J. Thermophys.* **2013**, *34*, 2343–2350. [CrossRef]
11. Woodcraft, A.L. Predicting the thermal conductivity of aluminium alloys in the cryogenic to room temperature range. *Cryogenics* **2005**, *45*, 421–431. [CrossRef]
12. Peet, M.J.; Hasan, H.S.; Bhadeshia, H.K.D.H. Prediction of thermal conductivity of steel. *Int. J. Heat Mass Transf.* **2011**, *54*, 2602–2608. [CrossRef]
13. Li, J.L.; Wang, X.X.; Zhang, N.; Wu, D.; Chen, R.S. Ductility drop of the solutionized Mg-Gd-Y-Zr alloy during tensile deformation at 350 °C. *J. Alloys Compd.* **2017**, *714*, 104–113. [CrossRef]
14. Peng, R.; Xu, C.; Li, Y.; Zhong, S.; Cao, X.; Ding, Y. Multiple-twinning induced recrystallization and texture optimization in a differential-temperature-rolled AZ31B magnesium alloy with excellent ductility. *Mater. Res. Lett.* **2022**, *10*, 318–326. [CrossRef]
15. Kada, S.R.; Lynch, P.A.; Kimpton, J.A.; Barnett, M.R. In-situ X-ray diffraction studies of slip and twinning in the presence of precipitates in AZ91 alloy. *Acta Mater.* **2016**, *119*, 145–156. [CrossRef]
16. Al-Samman, T.; Gottstein, G. Dynamic recrystallization during high temperature deformation of magnesium. *Mater. Sci. Eng. A* **2008**, *490*, 411–420. [CrossRef]
17. Zhang, H.; Li, Y.X.; Zhu, G.M.; Zhu, Q.C.; Qi, X.X.; Zheng, D.F.; Zeng, X.Q. Crystallographic features of <11$\bar{0}$00> axis tilt boundaries in a high strain rate deformed Mg-9Y alloy. *Mater. Charact.* **2021**, *182*, 111522. [CrossRef]
18. Jalali, M.S.; Zarei-Hanzaki, A.; Mosayebi, M.; Abedi, H.R.; Malekan, M.; Kahnooji, M.; Farabi, E.; Kim, S.-H. Unveiling the influence of dendrite characteristics on the slip/twinning activity and the strain hardening capacity of Mg-Sn-Li-Zn cast alloys. *J. Magnes. Alloys* **2022**. [CrossRef]
19. Hong, L.; Wang, R.; Zhang, X. Effects of Nd on microstructure and mechanical properties of as-cast Mg-12Gd-2Zn-xNd-0.4Zr alloys with stacking faults. *Int. J. Miner. Metall. Mater.* **2022**, *29*, 1570–1577. [CrossRef]

20. Jiang, L.; Jonas, J.J.; Luo, A.A.; Sachdev, A.K.; Godet, S. Influence of {10-12} extension twinning on the flow behavior of AZ31 Mg alloy. *Mater. Sci. Eng. A* **2007**, *445–446*, 302–309. [CrossRef]
21. Li, L.; Liu, W.; Qi, F.; Wu, D.; Zhang, Z. Effects of deformation twins on microstructure evolution, mechanical properties and corrosion behaviors in magnesium alloys—A review. *J. Magnes. Alloys* **2022**, *10*, 2334–2353. [CrossRef]
22. Stanford, N.; Marceau, R.K.W.; Barnett, M.R. The effect of high yttrium solute concentration on the twinning behaviour of magnesium alloys. *Acta Mater.* **2015**, *82*, 447–456. [CrossRef]
23. Wang, B.; Wang, F.; Wang, Z.; Zhou, L.; Liu, Z.; Mao, P. Compressive deformation behavior of ultrafine-grained Mg-3Zn-1.2Ca-0.6Zr alloy at room temperature. *J. Alloys Compd.* **2021**, *871*, 159581. [CrossRef]
24. Tahaghoghi, M.; Zarei-Hanzaki, A.; Jalali, M.S.; Abedi, H.R. Improved strength and plasticity of magnesium matrix nanocomposites reinforced by carbonaceous nanoplatelets and micro-clusters. *J. Mater. Res. Technol.* **2022**, *21*, 2797–2814. [CrossRef]
25. Wang, B.; Xin, R.; Huang, G.; Liu, Q. Effect of crystal orientation on the mechanical properties and strain hardening behavior of magnesium alloy AZ31 during uniaxial compression. *Mater. Sci. Eng. A* **2012**, *534*, 588–593. [CrossRef]
26. Barnett, M.R.; Ghaderi, A.; Quinta da Fonseca, J.; Robson, J.D. Influence of orientation on twin nucleation and growth at low strains in a magnesium alloy. *Acta Mater.* **2014**, *80*, 380–391. [CrossRef]
27. Tao, J.-Q.; Cheng, Y.-S.; Huang, S.-D.; Peng, F.-F.; Yang, W.-X.; Lu, M.-Q.; Zhang, Z.-M.; Jin, X. Microstructural evolution and mechanical properties of ZK60 magnesium alloy prepared by multi-axial forging during partial remelting. *Trans. Nonferrous Met. Soc. China* **2012**, *22* (Suppl. S2), 428–434. [CrossRef]
28. Chen, H.; Zang, Q.; Yu, H.; Zhang, J.; Jin, Y. Effect of intermediate annealing on the microstructure and mechanical property of ZK60 magnesium alloy produced by twin roll casting and hot rolling. *Mater. Charact.* **2015**, *106*, 437–441. [CrossRef]
29. Fu, W.; Wang, R.; Zhang, J.; Wu, K.; Liu, G.; Sun, J. The effect of precipitates on voiding, twinning, and fracture behaviors in Mg alloys. *Mater. Sci. Eng. A* **2018**, *720*, 98–109. [CrossRef]
30. Robson, J.D.; Stanford, N.; Barnett, M.R. Effect of precipitate shape on slip and twinning in magnesium alloys. *Acta Mater.* **2011**, *59*, 1945–1956. [CrossRef]
31. Fan, H.; Aubry, S.; Arsenlis, A.; El-Awady, J.A. The role of twinning deformation on the hardening response of polycrystalline magnesium from discrete dislocation dynamics simulations. *Acta Mater.* **2015**, *92*, 126–139. [CrossRef]
32. Jiang, L.; Jonas, J.J.; Mishra, R.K.; Luo, A.A.; Sachdev, A.K.; Godet, S. Twinning and texture development in two Mg alloys subjected to loading along three different strain paths. *Acta Mater.* **2007**, *55*, 3899–3910. [CrossRef]
33. Koike, J. Enhanced deformation mechanisms by anisotropic plasticity in polycrystalline Mg alloys at room temperature. *Metall. Mater. Trans. A* **2005**, *36*, 1689–1696. [CrossRef]
34. Lu, D.; Wang, S.; Lan, Y.; Zhang, K.; Li, W.; Li, Q. Statistical Analysis of Grain-Scale Effects of Twinning Deformation for Magnesium Alloys under Cyclic Strain Loading. *Materials* **2020**, *13*, 2454. [CrossRef] [PubMed]
35. Chen, P.; Li, B.; Culbertson, D.; Jiang, Y. Contribution of extension twinning to plastic strain at low stress stage deformation of a Mg-3Al-1Zn alloy. *Mater. Sci. Eng. A* **2018**, *709*, 40–45. [CrossRef]
36. Chapuis, A.; Driver, J.H. Temperature dependency of slip and twinning in plane strain compressed magnesium single crystals. *Acta Mater.* **2011**, *59*, 1986–1994. [CrossRef]
37. El Kadiri, H.; Kapil, J.; Oppedal, A.L.; Hector, L.G.; Agnew, S.R.; Cherkaoui, M.; Vogel, S.C. The effect of twin–twin interactions on the nucleation and propagation of {101¯2} twinning in magnesium. *Acta Mater.* **2013**, *61*, 3549–3563. [CrossRef]
38. Gui, Y.; Li, Q.; Xue, Y.; Ouyang, L. Twin-twin geometric structure effect on the twinning behavior of an Mg-4Y-3Nd-2Sm-0.5Zr alloy traced by quasi-in-situ EBSD. *J. Magnes. Alloys* **2021**, in press. [CrossRef]
39. Mokdad, F.; Chen, D.L.; Li, D.Y. Single and double twin nucleation, growth, and interaction in an extruded magnesium alloy. *Mater. Des.* **2017**, *119*, 376–396. [CrossRef]
40. Gui, Y.; Cui, Y.; Bian, H.; Li, Q.; Ouyang, L.; Chiba, A. Role of slip and {10-12} twin on the crystal plasticity in Mg-RE alloy during deformation process at room temperature. *J. Mater. Sci. Technol.* **2021**, *80*, 279–296. [CrossRef]
41. Robson, J.D.; Stanford, N.; Barnett, M.R. Effect of particles in promoting twin nucleation in a Mg-5 wt.% Zn alloy. *Scr. Mater.* **2010**, *63*, 823–826. [CrossRef]
42. Drozdenko, D.; Dobroň, P.; Yi, S.; Horváth, K.; Letzig, D.; Bohlen, J. Mobility of pinned twin boundaries during mechanical loading of extruded binary Mg-1Zn alloy. *Mater. Charact.* **2018**, *139*, 81–88. [CrossRef]
43. Zhao, C.; Li, Z.; Shi, J.; Chen, X.; Tu, T.; Luo, Z.; Cheng, R.; Atrens, A.; Pan, F. Strain hardening behavior of Mg-Y alloys after extrusion process. *J. Magnes. Alloys* **2019**, *7*, 672–680. [CrossRef]
44. Liu, T.; Pan, F.; Zhang, X. Effect of Sc addition on the work-hardening behavior of ZK60 magnesium alloy. *Mater. Des.* **2013**, *43*, 572–577. [CrossRef]
45. Ang, H.Q. Modelling of the Strain Hardening Behaviour of Die-Cast Magnesium-Aluminium-Rare Earth Alloy. *Adv. Eng. Forum* **2020**, *35*, 1–8. [CrossRef]
46. Tong, L.B.; Zheng, M.Y.; Xu, S.W.; Hu, X.S.; Wu, K.; Kamado, S.; Wang, G.J.; Lv, X.Y. Room-temperature compressive deformation behavior of Mg–Zn–Ca alloy processed by equal channel angular pressing. *Mater. Sci. Eng. A* **2010**, *528*, 672–679. [CrossRef]

Disclaimer/Publisher's Note: The statements, opinions and data contained in all publications are solely those of the individual author(s) and contributor(s) and not of MDPI and/or the editor(s). MDPI and/or the editor(s) disclaim responsibility for any injury to people or property resulting from any ideas, methods, instructions or products referred to in the content.

Article

Achieving High-Strength and Toughness in a Mg-Gd-Y Alloy Using Multidirectional Impact Forging

Songhe Lu [1,2,†], Di Wu [2,†], Ming Yan [1] and Rongshi Chen [2,*]

1. Academy for Advanced Interdisciplinary Studies, Southern University of Science and Technology, Shenzhen 518055, China; lush3@sustech.edu.cn (S.L.); yanm@sustech.edu.cn (M.Y.)
2. Shi-Changxu Innovation Center for Advanced Materials, Institute of Metal Research, Chinese Academy of Sciences, 72 Wenhua Road, Shenyang 110016, China; dwu@imr.ac.cn
* Correspondence: rschen@imr.ac.cn; Tel.: +86-24-2392-6646; Fax: +86-24-2389-4149
† These authors contributed equally to this work.

Abstract: High strength and toughness are achieved in the Mg-4.96Gd-2.44Y-0.43Zr alloy by multidirectional impact forging (MDIF). The forged sample has a fine-grained microstructure with an average grain size of ~5.7 μm and a weak non-basal texture, and it was characterized by an optical microscope (OM), scanning electron microscope (SEM), and electron back-scattering diffraction (EBSD). Tensile results exhibit the tensile yield strength (TYS) and static toughness (ST) of as-homogenized alloy dramatically increased after forging and aging, i.e., the TYS increased from 135^{+4}_{-5} MPa to 337^{+2}_{-2} MPa, and the ST enhanced from $22.0^{+0.3}_{-0.5}$ MJ/m^3 to $50.4^{+5.3}_{-5.4}$ MJ/m^3. Specifically, the forged Mg-Gd-Y-Zr alloy owns higher TYS than that of commercial rolled WE54 (Mg-5.25Y-3.5Nd-0.5Zr) and WE43 (Mg-4.0Y-3.0Nd-0.5Zr) alloys.

Keywords: grain refinement; high strength and toughness; {10–12} twin; Mg-Gd-Y alloy; multidirectional impact forging

1. Introduction

Mg and its alloys, as structural materials for automobile and electronics industries, can meet the demands of weight reduction and increasing vehicle efficiency [1,2]. Unluckily, they exhibited relatively low yield strength and toughness in comparison to their competitors Al and Ti alloys [3]. To improve their mechanical property, some strengthening mechanisms (including solution and precipitation strengthening) have been applied to block the dislocation glide and achieve the enhancement of the yield strength but the degradation of the toughness [4–6]. In contrast, fine-grained strengthening has been considered as a promising way for synchronously enhancing the strength and toughness of Mg alloys. Furthermore, the corresponding strengthening effect in Mg alloys is more remarkable due to their larger slope coefficient of the Hall-Petch relationship (~3 times larger than that of aluminum alloy) [7].

Fine-grained AZ31 [8], AZ61 [9], AZ80 alloys [10] have been produced mainly by conventional deformation routes including rolling, extrusion, and forging, and the corresponding recrystallization predominantly proceeded from prior grain boundaries and then consumed the entire deformed microstructure easily. However, the above recrystallization mechanism was seriously suppressed in some RE (rare earth)-containing Mg alloys [11,12]. For instance, the microstructure of Mg-8.2Gd-3.8Y-1.0Zn-0.4Zr (wt.%) alloy rolled at 400 °C consisted of a large volume of residual coarse grains and some recrystallization distributed at the prior grain boundary [11]. A similar recrystallization behavior also was witnessed in the microstructure of extruded Mg-7.5Gd-2.5Y-3.5Zn-0.9Ca-0.4Zr (wt.%) alloy [12]. This should be originated from the strong solute drag or pinning effect of RE-rich precipitates on dislocation glide and rearrangement, especially for those grain boundaries having intensive segregation of solute atoms. In other words, these grain

Citation: Lu, S.; Wu, D.; Yan, M.; Chen, R. Achieving High-Strength and Toughness in a Mg-Gd-Y Alloy Using Multidirectional Impact Forging. *Materials* **2022**, *15*, 1508. https://doi.org/10.3390/ma15041508

Academic Editor: Frank Czerwinski

Received: 10 January 2022
Accepted: 27 January 2022
Published: 17 February 2022

Publisher's Note: MDPI stays neutral with regard to jurisdictional claims in published maps and institutional affiliations.

Copyright: © 2022 by the authors. Licensee MDPI, Basel, Switzerland. This article is an open access article distributed under the terms and conditions of the Creative Commons Attribution (CC BY) license (https://creativecommons.org/licenses/by/4.0/).

boundaries have already lost the function of recrystallization nucleation. To produce the fine-grained RE-containing Mg alloys, it is necessary to offer appropriate nucleation sites in the grain interior for recrystallization.

Mechanical twinning can serve as an important deformation mechanism in the plastic deformation of the Mg-RE alloys [13]. The corresponding twin boundaries generated in the grain interior may be the preliminary candidate for recrystallization nucleation [14]. For common {10–12} extension, {10–11} contraction and {10–11}–{10–12} double twins, the latter two is difficulty to be utilized well due to their larger critical resolved shear stress (CRSS $\geq$ 70 MPa) [15,16] and crack tendency [17]. In contrast, the former {10–12} twin with relatively low CRSS (~10 MPa) is easy to be activated and accommodates the extension strain along the c-axis [18]. Hence, the {10–12} twin boundary may take the place of the initial grain boundary and act as preference nucleation sites for recrystallization of the RE-containing Mg alloys.

Recently, {10–12} twin activated by multidirectional impact forging (MDIF) has been successfully applied to activate {10–12} twin and produce fine-grained Mg-6.68Gd-5.9Y-0.48Zr (wt.%) alloy [19,20]. In contrast to recrystallization assisted by dislocation rearrangement, {10–12} twin induced recrystallization has many advantages as follows. Firstly, mutual intersections of two-dimensional {10–12} twin boundaries are easy to form three-dimensional recrystallization nuclei. Secondly, the fresh twin boundary can store enough strain energy to promote recrystallization nucleation through the repeated twin-dislocation interaction during the dynamic forging procedure [19]. Finally, recrystallization grains initiated from {10–12} twin boundary are more effective in achieving microstructure refinement of the Mg-Gd-Y alloy alloys.

In the present work, a high-strength and toughness Mg-Gd-Y alloy has been developed by MDIF using {10–12} twin and correlated recrystallization. Then, the contribution of {10–12} twins to microstructure refinement and property enhancement of Mg-Gd-Y-Zr alloy have been discussed in detail.

2. Materials and Methods

The material used in the present study was GW52 alloy ingot (Mg-4.96Gd-2.44Y-0.43Zr, mass fraction, %). The as-cast GW52 alloy ingot was homogenized at 480 °C for 8 h with subsequent cooling in air, and then it was cut into some cubic sample with dimensions of 70 × 70 × 70 mm^3. These cubic samples were first heated to 480 °C in an electric resistance furnace and kept for 60 min. In addition, the heating rate is 15 °C/min. Then, the multidirectional impact forging (MDIF) was carried out using an industrial air pneumatic hammer machine (Shandong Chu Hang Heavy Industry Machinery Co., Ltd., Weifang, China) with a load gravity of 250 Kg. A pass strain ~0.05 was applied and the average strain rate was around 20 s^{-1}. The procedure during the MDIF was shown in Figure 1a. One of the cubic samples was forged to 200 passes with an intermediate annealing treatment for ~10 min at 480 °C, and it was termed as MDIF100+100 sample (symbol + means an intermediate annealing treatment for ~10 min after the first 100 forging passes). The entire forging and annealing process finished in ~15 min. After carefully checking, the MDIF100+100 sample was free from any surface defects, as shown in its macro-morphology image Figure 1b.

To examine the microstructure and mechanical property, the MDIF100+100 sample was sectioned in the center along the last forging direction (LFD), as shown in Figure 1c. The corresponding specimens for optical microscopy (OM) (Carl Zeiss AG, Oberkochen, Germany), scanning electron microscopy (SEM) (Carl Zeiss AG, Oberkochen, Germany), and tension test was machined from the center part, as indicated by red arrows in Figure 1c. The electron back-scattering diffraction (EBSD) observations were carried out using a TESCAN-MIRA3 SEM (TESCAN, Brno, Czech Republic) operating at 20 kV and applying a corresponding probe current of 60 nA. Orientation imaging microscopy was measured at a step size of 1 μm and the acquired EBSD data was analyzed using the software of Aztec Crystal 4.3 (Oxford Instruments, Abingdon, UK). Texture analysis was conducted using the Schultz reflection method through X-ray diffraction and calculated pole figures were obtained using the DIFFRAC plus TEXEVAI software (Bruker, Karlsruhe, Germany). The tensile specimens in dog bone shape were cut from the central part of the forged cubic sample with a gauge length of 20 mm, a width of 3 mm, and a thickness of 2.5 mm. The compression specimens in-cylinder were also machined from the same position with a height of 15 mm and a diameter of 10 mm, as shown in Figure 1c.

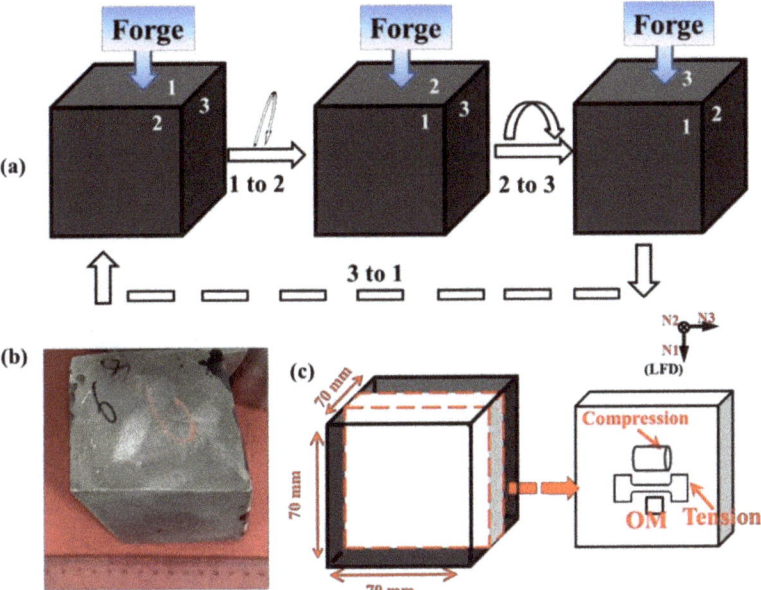

Figure 1. (a) Schematic illustrations of the MDIF process, (b) macro-morphology of MDIF100+100 sample and (c) schematic diagram describing the orientations of the tensile specimens and the compression specimens relative to the forged billet.

3. Results

3.1. Initial State of the Alloy

Figure 2 displays the EBSD results of the microstructure of GW52 alloy after homogenization at 480 °C for 8 h. The inverse pole figure exhibit many equiaxed grains with random orientation in the microstructure of as-homogenized alloy, as shown in Figure 2a. This is consistent with the {0001} pole figure in Figure 2d. The corresponding grain boundary map consists of an as-homogenized microstructure with some second phases in Figure 2b. Most of these phases in Mg-Gd-Y were identified as cuboid-shaped phases in previous reference [20], which were detected as YH_2 through a combined analysis of secondary ion mass spectrometry (SIMS) and X-ray tomography (XRT). Many black dots in Figure 2b may be the corrosion products of the second phase produced by electrolytic polishing. The average grain size of 47 μm has been evaluated through the grain size distribution map in Figure 2c.

3.2. Microstructure and Mechanical Property of MDIF100+100 Sample

Figure 3 gives OM and corresponding {0001} macro texture of the MDIF100+100 sample of GW52 alloy. It can be observed a deformed microstructure in low magnification of OM in Figure 3a. High-magnification observation captured some coarse grains ~10 μm and a large number of small grains ~1 μm, as indicated by their respective arrows in Figure 3b. Besides, the corresponding {0001} macrotexture in Figure 3c illustrates a non-basal texture with many peaks (the intensity ≤ 3.79 multiples of random (mrd)).

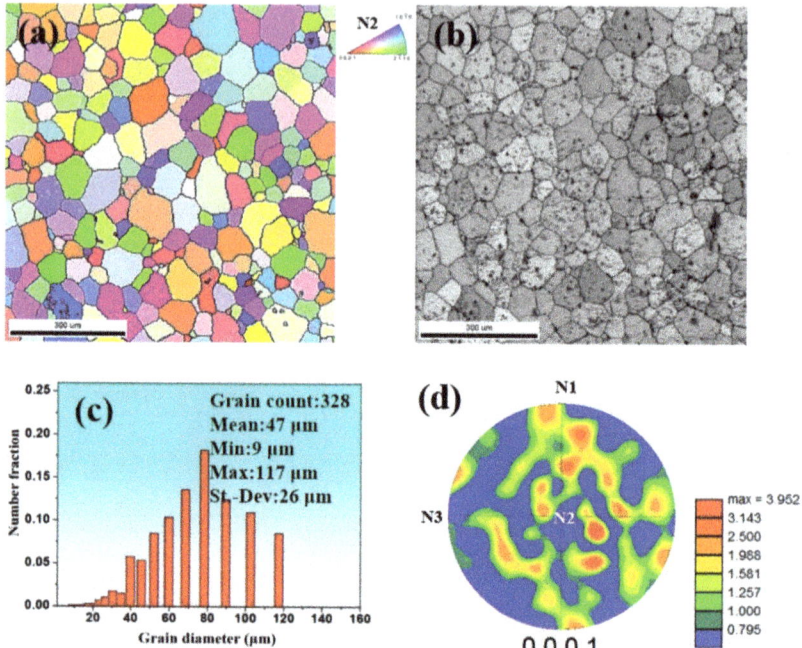

Figure 2. (a) Inverse pole figure, (b) grain boundary image, (c) grain size distribution map, and (d) {0001} pole figure of as-homogenized GW52 alloy.

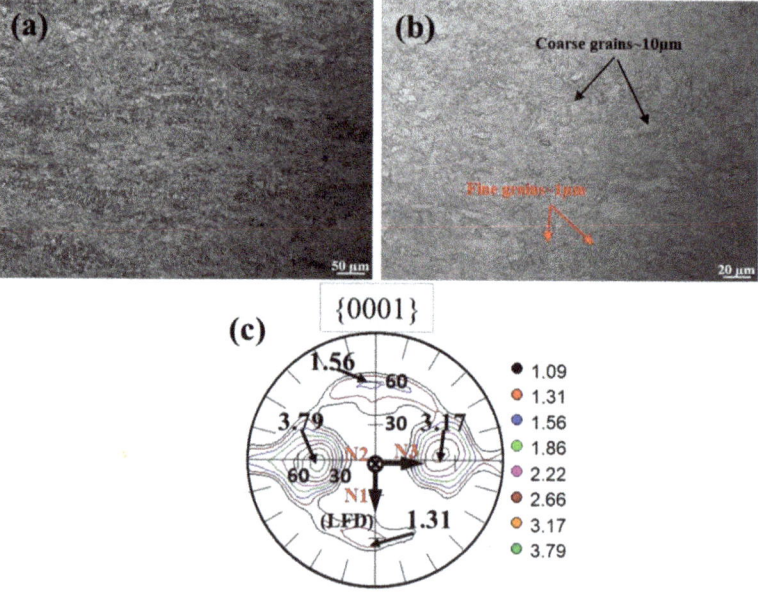

Figure 3. (a) Low magnification and (b) high magnification of OM as well as corresponding (c) {0001} macrotexture of MDIF100+100 sample of GW52 Mg alloy at the center location (~35 mm).

The corresponding EBSD results of the final forging microstructure were shown in Figure 4a–e. We noted that flourishing twins were activated in the microstructure, especially in several coarse regions, as shown in Figure 4a,b. The special boundaries results in Figure 4b demonstrate that the fraction of {10–12} twin boundary (<11–20> 86°)is 11.8%, while the fraction of {10–11} twin boundary (<11–20> 56°) and {10–11}–{10–12} twin boundary <11–20> 38° are 0.23% and 0.40%, respectively. This is consistent with the larger fraction of {10–12} twin boundaries in the misorientation angle of grain boundaries in Figure 4e. In addition, low angle grain boundaries (LAGBs) also have a larger number fraction in this forged alloy in Figure 4e. As shown in the grain size distribution histogram of Figure 4d, the final average grain diameter of the GW52 alloy was decreased to ~5.7 μm. Moreover, the fraction of the fine grins (≤20 μm) in the final forging microstructure was about 87%. In particular, the {0001} pole figure in Figure 4d exhibited the fine-grained alloy possess a non-basal texture (the intensity ≤ 4.13 mrd). Specifically, MDIF with intermediate annealing treatment is an effective route to produce fine-grained Mg-RE alloys with non-basal texture.

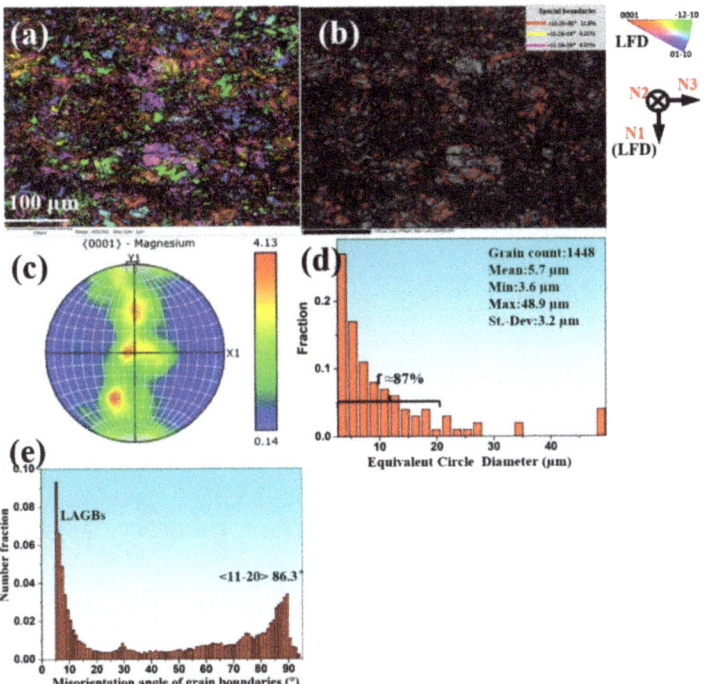

Figure 4. (a) Inverse pole figure map, (b) band contrast map, (c) {0001} pole figure, (d) grain size distribution histogram, and (e) misorientation angle of grain boundaries of MDIF100+100 sample of GW52 alloy at the center location (~35 mm).

Figure 5 presents room temperature tensile engineering stress-strain curves of as-homogenized, forged, and aged GW52 alloy. The tension results were listed in Table 1. The static toughness (ST) usually can be measured by the true stress-strain curve, which can be obtained from their respective engineering stress-strain curve through data processing. The true stress can be calculated by this equation: $S = \sigma^* (1 + \varepsilon)$. Then, ST was measured by the equation $ST = \int_0^{\varepsilon_k} S d\varepsilon$ [21,22]. In addition, S means true stress and ε_k is the value of true strain after the sample fractured. The corresponding results have been illustrated in Table 1.

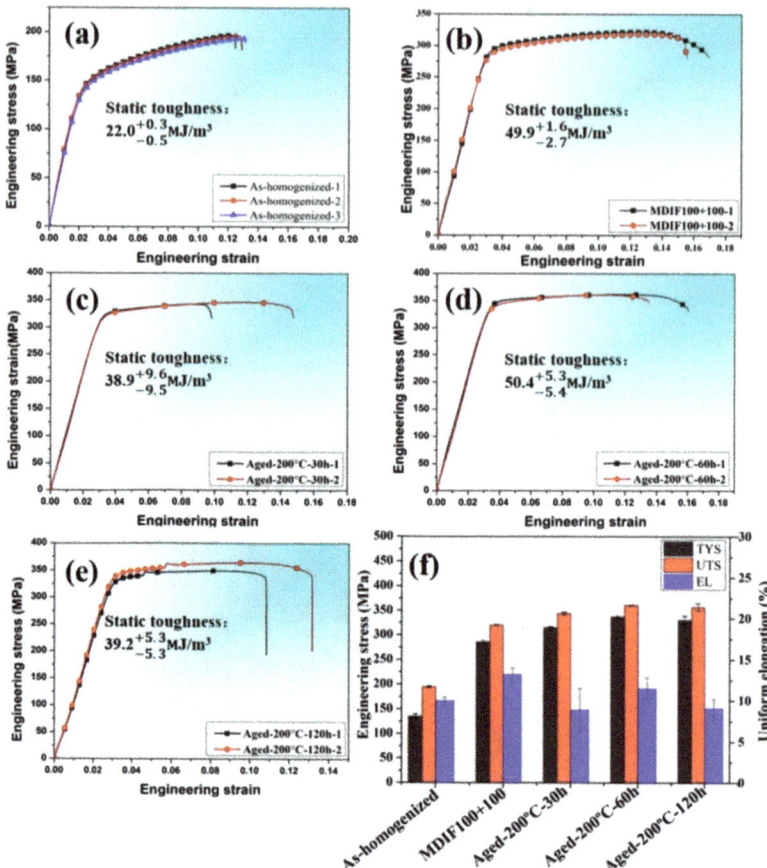

Figure 5. Room temperature tensile engineering stress-strain curves (**a–e**) and corresponding properties (**f**) of as-homogenized, MDIF100+100, and aged MDIF100+100 samples of GW52 alloy with their respective static toughness: (**a**) as-homogenized alloy, (**b**) MDIF100+100, (**c**) aged at 200 °C for 30 h, (**d**) aged at 200 °C for 60 h, (**e**) aged at 200 °C for 120 h.

Table 1. Room temperature tensile and compressive properties of GW52 alloy.

	TYS (MPa)	UT/CS(MPa)	EL (%)	ST(MJ/m^3)	CYS/TYS
As-homogenized-1	139	197	9.6	21.5	
As-homogenized-2	137	196	10.0	22.2	
As-homogenized-3	130	194	10.4	22.3	
MDIF100+100-ten-1	289	321	14.0	52.5	
MDIF100+100-ten-2	283	319	12.3	47.2	≈1
MDIF100+100-com-1	294	465	13.0	-	
MDIF100+100-com-2	310	469	11.0	-	
Aged-200 °C-30 h-1	316	341	6.2	29.3	
Aged-200 °C-30 h-2	313	346	11.5	48.4	
Aged-200 °C-60 h-1	339	361	12.8	55.7	
Aged-200 °C-60 h-2	335	360	10.2	45.0	
Aged-200 °C-120 h-1	322	364	7.9	33.9	
Aged-200 °C-120 h-2	339	349	10.2	44.5	

TYS, tension yield strength; UTS, ultimate tensile strength; EL, uniform elongation; CYS, compression yield strength; ST, static toughness.

It can be observed in Figure 5a that the tensile yield strength (TYS), uniform elongation (EL), and static toughness (ST) of as-homogenized GW52 alloy were 135^{+4}_{-5} MPa, $10.0^{+0.4}_{-0.4}$%, and $22.0^{+0.3}_{-0.5}$ MJ/m^3, respectively. After the MDIF of 100+100 passes, the strength, ductility, and toughness of GW52 alloy were synchronously enhanced in Figure 5b, i.e., the TYS, EL, and ST dramatically increased to 286^{+3}_{-3} MPa, $13.2^{+0.8}_{-0.9}$%, and $49.9^{+1.6}_{-1.7}$ MJ/m^3, respectively. It has been reported that aging treatment at 200 °C was usually applied to Mg-RE alloys to further enhance their strength [23–25]. In the present work, some tensile specimens of GW52 alloy were aged at 200 °C for different times (30 h, 60 h, and 120 h). The strength and toughness often were synchronously enhanced in Figure 5c–e, i.e., foraged alloy at 200 °C for 60 h, the TYS, UTS, and ST sharply increased to 337^{+2}_{-2} MPa, 361^{+0}_{-1} MPa, and $40.1^{+4.7}_{-5.3}$ MJ/m^3, respectively.

Figure 6a shows the comparison of mechanical properties in numerous forged Mg-Al-Zn alloys and Mg-Gd-Y-based alloys [10,26–39]. The forged Mg-Al-Zn-based alloys were generally located at the bottom corner [10,27–33], while the MDIF100+100 sample of GW52 alloy was located at the center part exhibiting a higher TYS. Furthermore, the forged GW52 alloy exhibited excellent strength–ductility balance in comparison to the forged Mg-Gd-Y-based alloys [34–39]. Specifically, the TYS and EL of the forged GW52 alloy are larger than that of rolling or extrusion of commercial WE54 and WE43 alloys produced by Magnesium Elektron Ltd. (Magnesium Elektron, Manchester, UK), in Figures S1 and S2 (Supplementary Materials), as shown in Figure 6a. Above all, the forged GW52 alloy demonstrates tension-compression yield symmetry along N3 forging direction (the ratio of CYS/TYS was ≈ 1.0) in Figure 6b, which means that we obtained yield isotropy Mg-Gd-Y alloy with high strength and toughness. This tension-compression yield symmetry character is rarely examined in the rolling sheet or extrusion of high-strength Mg-RE alloy [40]. It can be concluded that a fine-grained GW52 alloy has been produced by MDIF using {10–12} twin and correlated recrystallization with the excellent mechanical property. Thus, the contribution of {10–12} twins to microstructure refinement and property enhancement of GW52 alloy should be discussed in detail.

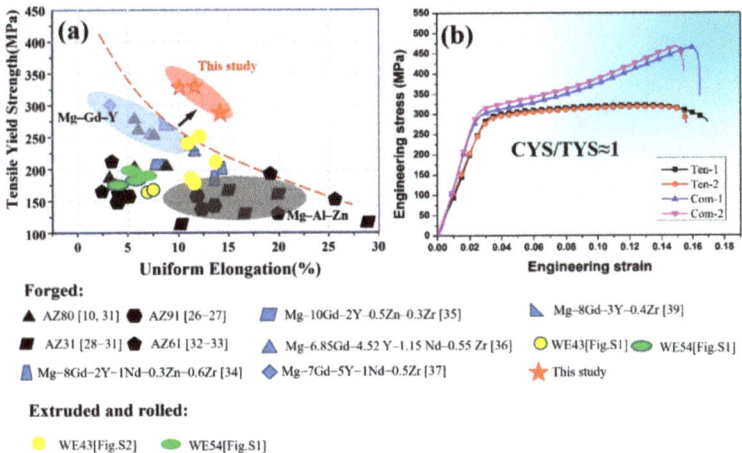

Figure 6. (a) Relationships between tensile yield strength and elongation of forged Mg-Al-Zn based and Mg-Gd-Y-based alloys as well as (b) tension-compression engineering stress-strain curves of MDIF100+100 sample of GW52 alloy.

4. Discussion

4.1. Contribution of {10–12} Twins to Microstructure Refinement

It has been reported that {10–12} twins usually exerted different functions at dynamic forging and annealing treatment in our previous works [19,20]. At the dynamic forging process, MDIF with a high strain rate and small strain is favorable for twin nucleation but twin growth [20]. Additionally, the three directional loadings are beneficial for activating {10–12} twins in 79.8% of grains in the microstructure of the as-homogenized Mg-Gd-Y alloy [20]. Meanwhile, these twins can divide their parent grains, interact repeatedly with various dislocations, and then store enough strain energy, eventually promoting the nucleation of DRX [20]. With the progress of intermediate annealing, above {10–12} twins with high strain energy can act as preference nucleation sites for static recrystallization, thereby making for a preliminary microstructure refinement [19]. Next, this fine-grained microstructure was ready for the next generation of {10–12} twin. At the sequent dynamic forging passes, the fresh {10–12} twins can be activated again and subdivide those new grains achieving further microstructure refinement. It can be concluded that MDIF with intermediate annealing can effectively refine the microstructure of Mg-Gd-Y alloy through {10–12} twin and correlated recrystallization.

The above microstructural results can be evidenced by the fine-grained microstructure with extensive {10–12} twins, as shown in inverse pole figure map and band contrast map of Figure 4a,b. The distribution histogram of grain size had been examined in Figure 4d, and the average grain size of the GW52 alloy sharply decreased from ~47 μm of as-homogenized alloy to ~5.7 μm. We noted that the fraction of the fine grins ($\leq$20 μm) dramatically increased to 87%. This means that MDIF with intermediate annealing had produced a fine-grained Mg-Gd-Y alloy using {10–12} twin and correlated recrystallization.

4.2. High Strength and Toughness GW52 Alloy

After the MDIF of 100+100 forging passes, the TYS of GW52 alloy increased from 135^{+4}_{-5} MPa to 286^{+3}_{-3} MPa and the ST also increased from $22.0^{+0.3}_{-0.5}$ MJ/m^3, to $49.9^{+1.6}_{-1.7}$ MJ/m^3. As a result, the MDIF realized 151 MPa increment (112%) for TYS and 22.0 MJ/m^3 increment (117%) for ST. Meanwhile, the average grain size of the alloy sharply decreased from ~47 μm of as-homogenized alloy to ~5.7 μm. The strengthening and toughening mechanisms will be discussed in detail as follows.

For the strengthening mechanism, fine-grained strengthening and twin strengthening should be focused on in the present work. Based on the stress intensity factor, k, obtained using the Hall–Petch relationship of GW53 alloy [41], the increment of TYS induced by grain refinement was calculated at about 112 MPa. It can be concluded that fine-grained strengthening determines the increment of TYS of GW52 alloy. Besides, flourishing twins were activated in the microstructure, especially in several coarse regions, as shown in Figure 4a,b. The special boundaries results in Figure 4b demonstrate that the fraction of {10–12} twin boundary (<11–20> 86°) is 11.8%, while the fraction of {10–11} twin boundary (<11–20> 56°) and {10–11}–{10–12} twin boundary <11–20> 38° are 0.23% and 0.40%, respectively. As a result, the crack tendency of {10–11} compression and {10–11}–{10–12} double twin can almost be neglected [42,43]. As a result, a large number of {10–12} twin boundaries as typical two-dimensional planar defects can interact with basal slip and even impede the dislocation movement [44]. Therefore, fine grain and twin strengthening play a key role in the high-strength GW52 alloy.

For the toughening mechanism, intergranular deformation including grain sliding and grain rotation usually serves as the dominating deformation mechanism in the plastic deformation of fine-grained GW52 alloy (~5.7 μm). Meanwhile, deformation mechanisms will have a significant transition from primary basal slip to some potential non-basal slip in fine-grained alloy with the grain size < 10 μm [45]. In addition, a large volume of {10–12} twin boundaries as typical two-dimensional planar defects exerted an obstructing effect on dislocation glide and allow the penetration of dislocation under the high stress-concentration [44]. On the other hand, the fine-grained GW52 alloy with a random texture tends to activate various dislocation slip, which was evidenced by their higher average SF (Schmid factor) value ($\geq$0.28) for various slip systems (basal slip <a>: {0001} <11–20>, prismatic slip <a>: {1–100} <11–20>, pyramidal slip <a>: {1–101} <11–20>, and pyramidal slip <c+a>: {11–22} <11-2-3>), as shown in Figure 7. In other words, various slips will be activated and steadily accommodate the tensile strain according to the Von Mises criteria [46]. All of these changes can ensure a uniform deformation and suppress the premature failure of Mg alloy. Hence, the fine-grained microstructure obtained by MDIF plays an important role in this high strength and toughness Mg-Gd-Y alloy.

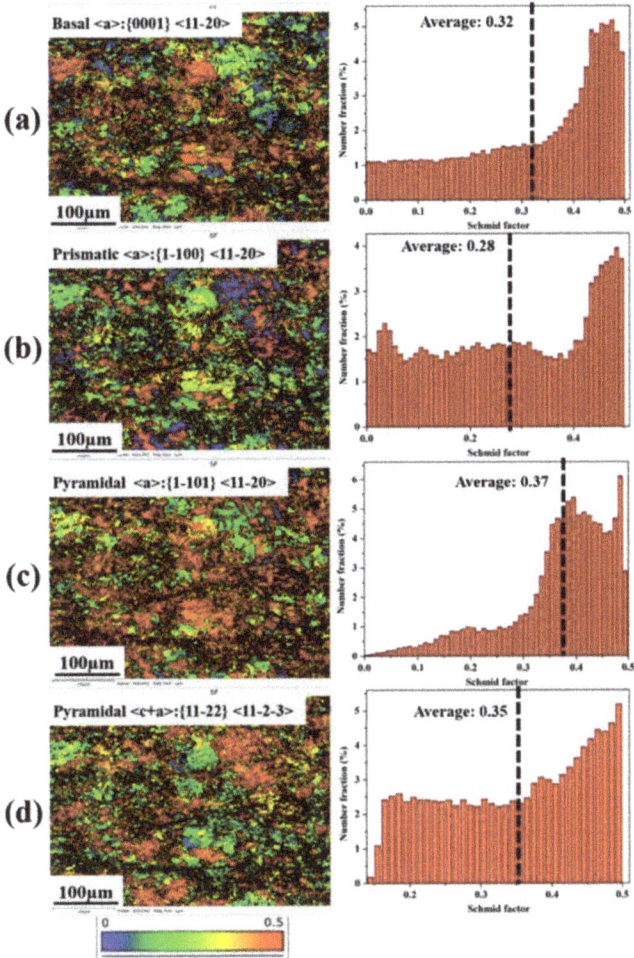

Figure 7. SF maps and corresponding distribution histograms for various slip systems of MDIF100+100 sample of GW52 alloy: (**a**) basal slip <a>:{0001} <11–20>, (**b**) prismatic slip <a>:{1–100} <11–20>, (**c**) pyramidal slip <a>:{1–101} <11–20>, (**d**) pyramidal slip <c+a>:{11–22} <11-2-3>.

5. Conclusions

Multidirectional impact forging has been applied to grain refinement and property enhancement of Mg-Gd-Y alloy in the present work. The main results can be summarized as follows:

(1) Multidirectional impact forging has been proved to be an efficient methodology in grain refinement and property improvement of Mg-Gd-Y alloy. After the MDIF of 100+100 forging passes, the TYS of GW52 alloy increased from 135^{+4}_{-5} MPa to 286^{+3}_{-3} MPa and the ST also increased from $22.0^{+0.3}_{-0.5}$ MJ/m^3, to $49.9^{+1.6}_{-1.7}$ MJ/m^3. As

(2) The MDIF100+100 sample of GW52 alloy has a relatively fine-grained microstructure ~5.7 μm, exhibiting a random texture. Furthermore, high TYS of 337^{+2}_{-2} MPa, EL of $11.5^{+1.3}_{-1.3}$%, and ST of $50.4^{+5.3}_{-5.4}$ MJ/m^3 were gained in the GW52 alloy developed by MDIF and aging treatment.

(3) The forged GW52 alloy exhibits yield isotropy (the ratio of compression yield strength/tensile yield strength along the forging direction is ≈1.0). Compared with forged Mg-Al-Zn and Mg-Gd-Y based alloys, the forged GW52 alloy exhibited excellent strength–ductility balance resulting from the combined effect of fine-grained strengthening, twin strengthening, and its non-basal texture.

As a result, the MDIF realized 151 MPa increment (112%) for TYS and 22.0 MJ/m^3 increment (117%) for ST simultaneously.

Supplementary Materials: The following are available online at https://www.mdpi.com/article/10.3390/ma15041508/s1. Figure S1. Room temperature tensile properties of extruded bars and forged billets of commercial WE43 and WE54 alloys produced by Elektron Ltd. (London, UK); Figure S2. Room temperature tensile properties of rolled plate and extruded bar of commercial WE43 alloys produced by Elektron Ltd (London, UK).

Author Contributions: Investigation, writing—review & editing, S.L.; supervision, review & editing, D.W. and R.C.; review & editing, M.Y. All authors have read and agreed to the published version of the manuscript.

Funding: This work was funded by the National Natural Science Foundation of China (NSFC) through Projects No. 52171055, the National Science and Technology Major Project of China through Project No. 2017ZX04014001, and the Natural Science Foundation of Liaoning province through Projects No. 20180550799.

Institutional Review Board Statement: Not applicable.

Informed Consent Statement: Not applicable.

Data Availability Statement: The data presented in this study are available upon request from the corresponding author. The data are not publicly available due to the requirements of related projects.

Conflicts of Interest: The authors declare they have no known competing financial interests or personal relationships that could have influenced the work reported in this paper.

References

1. Zhang, K.; Zheng, J.-H.; Huang, Y.; Pruncu, C.; Jiang, J. Evolution of twinning and shear bands in magnesium alloys during rolling at room and cryogenic temperature. *Mater. Des.* **2020**, *193*, 108793. [CrossRef]
2. Pan, H.; Kang, R.; Li, J.; Xie, H.; Zeng, Z.; Huang, Q.; Yang, C.; Ren, Y.; Qin, G. Mechanistic investigation of a low-alloy Mg–Ca-based extrusion alloy with high strength–ductility synergy. *Acta Mater.* **2020**, *186*, 278–290. [CrossRef]
3. Springer, H.; Baron, C.; Szczepaniak, A.; Uhlenwinkel, V.; Raabe, D. Stiff, light, strong and ductile: Nano-structured High Modulus Steel. *Sci. Rep.* **2017**, *7*, 2757. [CrossRef]
4. Zheng, R.; Du, J.-P.; Gao, S.; Somekawa, H.; Ogata, S.; Tsuji, N. Transition of dominant deformation mode in bulk polycrystalline pure Mg by ultra-grain refinement down to sub-micrometer. *Acta Mater.* **2020**, *198*, 35–46. [CrossRef]
5. Zhang, B.; Wang, Y.; Geng, L.; Lu, C. Effects of calcium on texture and mechanical properties of hot-extruded Mg-Zn-Ca alloys. *Mater. Sci. Eng. A* **2012**, *539*, 56–60. [CrossRef]
6. Hagihara, K.; Li, Z.; Yamasaki, M.; Kawamura, Y.; Nakano, T. Strengthening mechanisms acting in extruded Mg-based long-period stacking ordered (LPSO)-phase alloys. *Acta Mater.* **2019**, *163*, 226–239. [CrossRef]
7. Chen, Z.H. *Wrought Magnesiun*; Chemical Industry Press: Beijing, China, 2005; p. 5.
8. Jiang, M.; Xu, C.; Yan, H.; Fan, G.; Nakata, T.; Lao, C.; Chen, R.; Kamado, S.; Han, E.; Lu, B. Unveiling the formation of basal texture variations based on twinning and dynamic recrystallization in AZ31 magnesium alloy during extrusion. *Acta Mater.* **2018**, *157*, 53–71. [CrossRef]
9. Huang, X.; Suzuki, K.; Chino, Y.; Mabuchi, M. Texture and stretch formability of AZ61 and AM60 magnesium alloy sheets pro-cessed by high-temperature rolling. *J. Alloys Compd.* **2015**, *632*, 94–102. [CrossRef]
10. Zhou, X.; Zhang, J.; Chen, X.; Zhang, X.; Li, M. Fabrication of high-strength AZ80 alloys via multidirectional forging in air with no need of ageing treatment. *J. Alloys Compd.* **2019**, *787*, 551–559. [CrossRef]

11. Xu, C.; Zheng, M.Y.; Wu, K.; Wang, E.D.; Fan, G.H.; Xu, S.W.; Kamado, S.; Liu, X.D.; Wang, G.J.; Lv, X.Y.; et al. Effect of final rolling reduction on the microstructure and mechanical properties of Mg-Gd-Y-Zn-Zr alloy sheets. *Mater. Sci. Eng. A* **2013**, *559*, 232–240. [CrossRef]
12. Xie, Z.Y.; Tian, Y.; Li, Q.; Zhou, J.C.; Meng, Y. Effect of extrusion parameters on mi-crostructure and mechanical properties of Mg-7.5Gd-2.5Y-3.5Zn-0.9Ca-0.4Zr (wt%) alloy. *Mater. Sci. Eng. A* **2017**, *685*, 159–167.
13. Lu, S.; Wu, D.; Chen, R.; Han, E.-H. The influence of temperature on twinning behavior of a Mg-Gd-Y alloy during hot compression. *Mater. Sci. Eng. A* **2018**, *735*, 173–181. [CrossRef]
14. Lu, S.; Wu, D.; Chen, R.; Han, E.-H. The effect of twinning on dynamic recrystallization behavior of Mg-Gd-Y alloy during hot compression. *J. Alloys Compd.* **2019**, *803*, 277–290. [CrossRef]
15. Zhu, S.; Yan, H.; Liao, X.; Moody, S.; Sha, G.; Wu, Y.; Ringer, S. Mechanisms for enhanced plasticity in magnesium alloys. *Acta Mater.* **2015**, *82*, 344–355. [CrossRef]
16. Basu, I.; Al-Samman, T. Twin recrystallization mechanisms in magnesium-rare earth alloys. *Acta Mater.* **2015**, *96*, 111–132. [CrossRef]
17. Barnett, M. Twinning and the ductility of magnesium alloys: Part II. "Contraction" twins. *Mater. Sci. Eng. A* **2007**, *464*, 8–16. [CrossRef]
18. Chapuis, A.; Driver, J.H. Temperature dependency of slip and twinning in plane strain compressed magnesium single crystals. *Acta Mater.* **2011**, *59*, 1986–1994. [CrossRef]
19. Lu, S.H.; Wu, D.; Chen, R.S.; Han, E.H. Microstructure and texture optimization by static recrystallization originating from {10–12} extension twins in a Mg-Gd-Y alloy. *J. Mater. Sci. Technol.* **2020**, *59*, 44–60. [CrossRef]
20. Lu, S.H.; Wu, D.; Chen, R.S.; Han, E.H. Reasonable utilization of {10–12} twin for optimizing microstructure and improving mechanical property in a Mg-Gd-Y alloy. *Mater. Des.* **2020**, *191*, 108600–108621. [CrossRef]
21. Yao, T.Z.; Xu, T.H.; Wang, D.H. Effect of yield ratio rising on use security of pipeline steels. *Mater. Mech. Eng.* **2012**, *36*, 62–68.
22. Cai, W.; Morovat, M.A.; Engelhardt, M.D. True stress-strain curves for ASTM A992 steel for fracture simulation at elevated temperatures. *J. Constr. Steel Res.* **2017**, *139*, 272–279. [CrossRef]
23. Xue, Z.Y.; Ren, Y.J.; Luo, W.B.; Ren, Y.; Xu, P.; Xu, C. Microstructure and mechanical properties of Mg-Gd-Y-Zn-Zr alloy sheets processed by combined processes of extrusion, hot rolling and ageing. *Mater. Sci. Eng. A* **2013**, *559*, 844–851. [CrossRef]
24. Li, J.L.; Zhang, N.; Wang, X.X.; Wu, D.; Chen, R.S. Effect of Solution Treatment on the Microstructure and Mechanical Properties of Sand-Cast Mg-9Gd-4Y-0.5Zr Alloy. *Acta Metall. Sin. Engl. Lett.* **2017**, *31*, 189–198. [CrossRef]
25. Hou, X.; Cao, Z.; Wang, L.; Xu, S.; Kamado, S.; Wang, L. Microstructure and mechanical properties of extruded Mg-8Gd-2Y-1Nd-0.3Zn-0.6Zr alloy. *Mater. Sci. Eng. A* **2011**, *528*, 7805–7810. [CrossRef]
26. Nie, K.; Wang, X.; Deng, K.; Xu, F.; Wu, K.; Zheng, M. Microstructures and mechanical properties of AZ91 magnesium alloy processed by multidirectional forging under decreasing temperature conditions. *J. Alloys Compd.* **2014**, *617*, 979–987. [CrossRef]
27. Nie, K.; Deng, K.; Wang, X.; Xu, F.; Wu, K.; Zheng, M. Multidirectional forging of AZ91 magnesium alloy and its effects on microstructures and mechanical properties. *Mater. Sci. Eng. A* **2015**, *624*, 157–168. [CrossRef]
28. Guo, W.; Wang, Q.; Ye, B.; Zhou, H. Enhanced microstructure homogeneity and mechanical properties of AZ31–Si composite by cyclic closed-die forging. *J. Alloys Compd.* **2013**, *552*, 409–417. [CrossRef]
29. Liu, Z.Y.; Jiang, X.Q.; Mu, F.; Li, H.D. Development progress of plastic working for AZ31 magnesium alloy. *Light Met.* **2008**, *12*, 59–63. [CrossRef]
30. Chen, C.; Wang, R.; Du, X.H.; Wu, B.L. Improved Mechanical Properties of AZ31 Alloy Fabricated by Multi-Directional Forging. *Key Eng. Mater.* **2017**, *727*, 124–131. [CrossRef]
31. Guo, Q.; Yan, H.; Chen, Z.; Zhang, H. Effect of mutiple forging process on microstructure and mechanical properties of Magnesium alloy AZ80. *Acta Metall. Sin.* **2006**, *42*, 739–744.
32. Jiang, M.G.; Yan, H.; Chen, R.S. Microstructure, texture and mechanical properties in an as-cast AZ61 Mg alloy during multi-directional impact forging and subsequent heat treatment. *Mater. Des.* **2015**, *87*, 891–900. [CrossRef]
33. Jiang, M.; Yan, H.; Chen, R. Enhanced mechanical properties due to grain refinement and texture modification in an AZ61 Mg alloy processed by small strain impact forging. *Mater. Sci. Eng. A* **2015**, *621*, 204–211. [CrossRef]
34. Hou, X.L.; Cao, Z.Y.; Wang, L.D.; Wang, L.M. Investigation on the Microstructure and Mechanical Properties of Hot Forged Mg-8Gd-2Y-1Nd-0.3Zn-0.6Zr Alloy. *Adv. Mater. Res.* **2011**, *284–286*, 1598–1602. [CrossRef]
35. Han, X.Z.; Xu, W.C.; Shan, D.B. Effect of precipitates on microstructures and properties of forged Mg-10Gd-2Y-0.5Zn-0.3Zr alloy during ageing process. *J. Alloys Compd.* **2011**, *509*, 8625–8631. [CrossRef]
36. Xia, X.; Chen, Q.; Zhao, Z.; Ma, M.; Li, X.; Zhang, K. Microstructure, texture and mechanical properties of coarse-grained Mg-Gd-Y-Nd-Zr alloy processed by multidirectional forging. *J. Alloys Compd.* **2015**, *623*, 62–68. [CrossRef]
37. Wu, D.; Li, S.Q.; Hong, M.; Chen, R.S.; Han, E.H.; Ke, W. High cycle fatigue behavior of the forged Mg-7Gd-5Y-1Nd-0.5Zr alloy. *J. Magnes. Alloys* **2014**, *2*, 357–362. [CrossRef]
38. Dong, B.; Zhang, Z.; Yu, J.; Che, X.; Meng, M.; Zhang, J. Microstructure, texture evolution and mechanical properties of multi-directional forged Mg-13Gd-4Y-2Zn-0.5Zr alloy under decreasing temperature. *J. Alloys Compd.* **2020**, *823*, 153776. [CrossRef]
39. Liu, B.; Zhang, Z.; Jin, L.; Gao, J.; Dong, J. Forgeability, microstructure and mechanical properties of a free-forged Mg-8Gd-3Y-0.4Zr alloy. *Mater. Sci. Eng. A* **2016**, *650*, 233–239. [CrossRef]

40. Xin, R.; Song, B.; Zeng, K.; Huang, G.; Liu, Q. Effect of aging precipitation on mechanical anisotropy of an extruded Mg-Y-Nd alloy. *Mater. Des.* **2012**, *34*, 384–388. [CrossRef]
41. Wang, Z.Q.; Chen, J.; Wei, T.; Yan, W. Effect of Zr content on Grain size and Solution Treatment of Mg-5Gd-3Y-xZr alloy. *J. Xian Technol. Univ.* **2015**, *35*, 317–321.
42. Niknejad, S.; Esmaeili, S.; Zhou, N.Y. The role of double twinning on transgranular fracture in magnesium AZ61 in a localized stress field. *Acta Mater.* **2016**, *102*, 1–16. [CrossRef]
43. Cizek, P.; Barnett, M.R. Characteristics of the contraction twins formed close to the fracture surface in Mg–3Al–1Zn alloy de-formed in tension. *Scr. Mater.* **2008**, *59*, 959–962. [CrossRef]
44. Yu, H.; Xin, Y.; Chapuis, A.; Huang, X.; Xin, R.; Liu, Q. The different effects of twin boundary and grain boundary on reducing tension-compression yield asymmetry of Mg alloys. *Sci. Rep.* **2016**, *6*, 29283. [CrossRef] [PubMed]
45. Zhang, W.; Yu, Y.; Zhang, X.; Chen, W.; Wang, E. Mechanical anisotropy improvement in ultrafine-grained ZK61 magnesium alloy rods fabricated by cyclic extrusion and compression. *Mater. Sci. Eng. A* **2014**, *600*, 181–187. [CrossRef]
46. Mises, R.V. Mechanik der plastischen Formänderung von Kristallen. *J. Appl. Math. Mech.* **1928**, *8*, 161–185. [CrossRef]

Article

Simultaneously Improving Ductility and Stretch Formability of Mg-3Y Sheet via High Temperature Cross-Rolling and Subsequent Short-Term Annealing

Yinyang Wang, Chen Liu *, Yu Fu, Yongdong Xu *, Zhiwen Shao, Xiaohu Chen and Xiurong Zhu

Ningbo Branch of Chinese Academy of Ordnance Science, Ningbo 315103, China; wangyinyang929@126.com (Y.W.); fuyuayu@126.com (Y.F.); lourry@163.com (Z.S.); xiaohuzss@126.com (X.C.); zxr0922@163.com (X.Z.)
* Correspondence: chenliunbbky@gmail.com (C.L.); ydxunbbky@gmail.com (Y.X.)

Abstract: In this work, Mg-3Y sheet was prepared by high temperature cross-rolling and subsequent short-term annealing. The effect of annealing on microstructure, texture, mechanical properties, and stretch formability of Mg-3Y sheet was primarily investigated. Micro-nano size coexistence of β-$Mg_{24}Y_5$ phases can be well deformed with matrix. The as-rolled Mg-3Y sheet exhibited a homogeneous deformation microstructure consisting of deformed grains with extensive kink bands and dispersed β-$Mg_{24}Y_5$ phases. A double peak texture character appeared in as-rolled Mg-3Y sheet with a split of the texture peaks of about ±20° tilted to rolling direction. After annealing, the as-annealed Mg-3Y sheet presented complete static recrystallized (SRXed) microstructure consisting of uniform equiaxed grains. The texture orientation distribution was more dispersed and a weakened multiple-peak texture orientation distribution appeared. In addition, the maximum intensity of basal plane decreased from 5.2 to 3.1. The change of texture character was attributed to static recrystallization (SRX) induced by kink bands and grain boundaries. The as-annealed Mg-3Y sheet with high Schmid factor (SF) for basal <a> slip, prismatic <a> slip, pyramidal <a> slip, and pyramidal <c+a> slip exhibited high ductility (~25.6%). Simultaneously, enhanced activity of basal <a> slip and randomized grain orientation played a significant role in decreasing anisotropy for the as-annealed Mg-3Y sheet, which contributed to the formation of high stretch formability (~6.2 mm) at room temperature.

Keywords: microstructure; texture; mechanical properties; stretch formability; high temperature cross-rolling; annealing

Citation: Wang, Y.; Liu, C.; Fu, Y.; Xu, Y.; Shao, Z.; Chen, X.; Zhu, X. Simultaneously Improving Ductility and Stretch Formability of Mg-3Y Sheet via High Temperature Cross-Rolling and Subsequent Short-Term Annealing. *Materials* **2022**, *15*, 4712. https://doi.org/10.3390/ma15134712

Academic Editor: Jan Haubrich

Received: 23 May 2022
Accepted: 28 June 2022
Published: 5 July 2022

Publisher's Note: MDPI stays neutral with regard to jurisdictional claims in published maps and institutional affiliations.

Copyright: © 2022 by the authors. Licensee MDPI, Basel, Switzerland. This article is an open access article distributed under the terms and conditions of the Creative Commons Attribution (CC BY) license (https://creativecommons.org/licenses/by/4.0/).

1. Introduction

As the lightest structural material, current magnesium (Mg) alloy sheets cannot meet the requirements of automotive vehicles due to poor room temperature formability [1]. The poor formability and limited ductility of Mg alloys at ambient temperature is ascribed to their hexagonal closed-packed crystal structure and the associated insufficient independent slip systems [2–5]. Under hot deformation conditions, more slip systems would be operated in addition to the basal slip system and new fine grains could be concurrently developed [3–6]. These can lead to a pronounced improvement in the ductility and plastic workability of Mg products at not only high temperatures, but also warm and cold temperatures. However, strong basal texture would develop in rolled Mg alloy sheets. The basal planes are mainly parallel to the RD-TD plane, where the RD and TD are rolling and transverse directions [7–9]. The basal slip and extension twinning during deformation play a dominant role in the development of basal texture. These two mechanisms align the c-axis of grains with the direction of compressive strain, i.e., in as-rolled sheets, c-axis parallel with the normal direction (ND). The strong basal texture is difficult to remove during subsequent annealing [10]. Therefore, it is necessary to find a way of weakening the texture in wrought Mg alloys.

One of the methods of texture control is the addition of rare-earth (RE) metals in the Mg alloys [11–14], which could modify and weaken the intensity of the basal texture of rolled Mg alloys. Another approach is the improvement of the plastic processing techniques. In recent years, some processing techniques, such as asymmetric rolling [15,16], cross-rolling [17], repeated unidirectional bending [18], repetitive bending [19], equal channel angular rolling [20], wavy roll forming [21], high temperature annealing before and after warm rolling [22], combination of high temperature and warm rolling [23,24], have been explored for texture control of Mg alloy sheets. Compared with those processing techniques, high temperature rolling exhibits different deformation characteristics, such as enhanced activities of non-basal slips and grain boundary sliding (GBS), which could effectively weaken the texture [25,26]. It was found that high temperature rolling and subsequent annealing can significantly improve the stretch formability [27,28] and deep drawability [29] of Mg alloys at room temperature. Some reports have also shown that basal texture of Mg alloys could be suppressed by cross-rolling [30–32]. Meanwhile, a decrease in anisotropy in cross-rolling sheet was reported [33,34].

As mentioned above, Mg alloys with RE elements subjected to high temperature rolling and cross-rolling have potential to achieve high ductility and high stretch formability at room temperature. Therefore, in the present study, the Mg alloys with Y addition are processed by combination of high temperature rolling and cross-rolling under different conditions (hot rolling condition and hot rolling-annealing condition). The pure Mg is served as the control sample. The microstructure, texture, mechanical properties, and stretch formability of Mg-Y alloys are investigated in detail.

2. Materials and Methods

Commercially pure Mg and Mg-3Y (wt%) master alloys were used to prepare the pure Mg and Mg-3Y (wt%) alloys. The fusion metallurgy was carried out in a mild steel crucible placed in an electric resistance furnace under an anti-oxidizing flux. After melting, the melt was cast into a steel mold at 993 K. The as-cast alloys were homogenized at 723 K for 24 h, then cooled down in the air. The alloys were machined into a rectangular shape with dimensions of 100 mm × 80 mm × 18 mm. Before hot rolling, the as-homogenized samples were preheated at 773 K for 30 min. Each rolling direction of cross-rolling changed at 90°, as shown in Figure 1. To reduce strain hardening during the rolling, an incremental duration per pass was adopted, as shown in Figure 2. The rolling samples were inter-pass annealing treated at the rolling temperature for 10 min after each pass. After a total reduction of 70%, the as-rolled sheets were cooled in the air. The recrystallization annealing treatment was conducted at 748 K for 15 min for the final as-rolled alloy sheet.

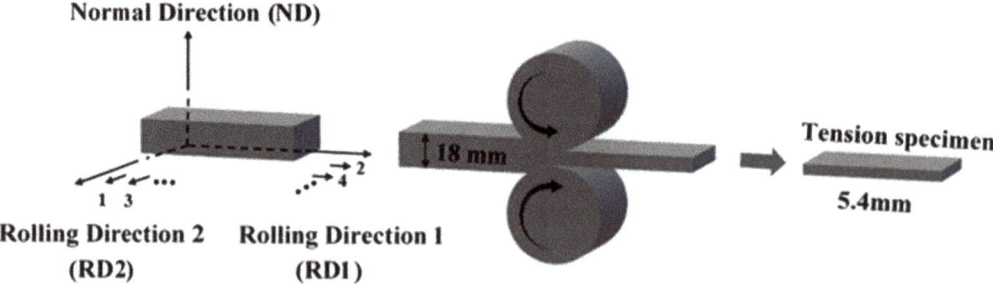

Figure 1. The schematic diagrams of hot cross-rolling.

The chemical compositions of alloys obtained by an X-ray fluorescence analyzer (XRF, XRF-1800) are listed in Table 1. The phase analysis and macro-texture test were performed by an X-ray diffraction (XRD, Empyrean) with Cu Kα radiation. The microstructure was observed by an optical microscope (OM, Leica MEF4) and a scanning electron microscope (SEM, SUPRA 55) equipped with energy dispersive X-ray spectroscopy (EDS). Electron backscattered diffraction (EBSD) analyses were carried out using a field emission scanning electron microscope (SEM, ZEISS EV55) equipped with an HKL EBSD detector. Substructure analysis was examined by transmission electron microscope (TEM, Tecnai G220 S-Twin). The specimens for the OM and SEM observation were prepared following a standard procedure of grinding, polishing, and etching (1.0 g picric acid+2 mL acetic acid+3 mL water+20 mL ethanol).

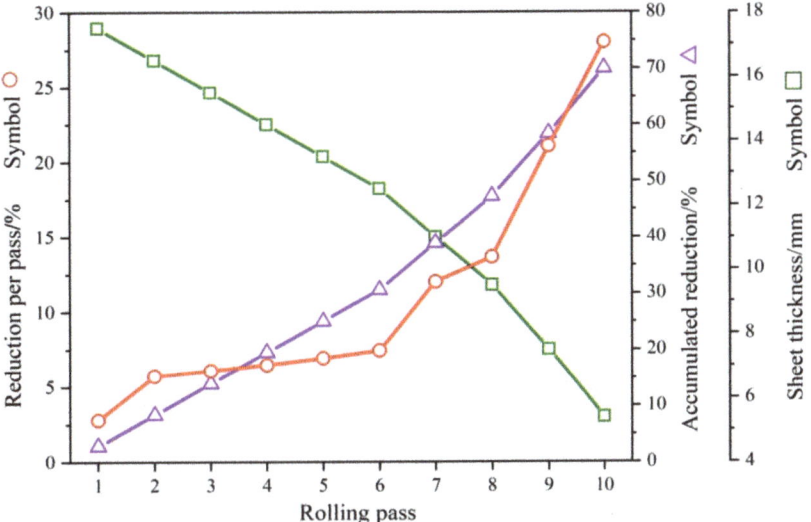

Figure 2. The reduction per pass, accumulated reduction, and sheet thickness as a function of rolling pass (red circle, blue triangle and green box represent reduction per pass, accumulated reduction and sheet thickness respectively).

Table 1. Chemical compositions of the as-cast materials (wt%).

Materials	Y	Mg
Mg	-	Bal.
Mg-3Y	2.90	Bal.

Tensile test was carried out using a DNS100 universal testing machine with a strain rate of 1×10^{-3} s^{-1} at room temperature. For each material, three parallel samples with their long axes parallel to the RD2 were tested. The fracture surface was also observed by the SUPRA 55 SEM. Erichsen test was carried out to determine the stretch formability of all parallel samples in each test using GBW-60Z Erichsen testing machine with a punch speed of 1 mm/min at room temperature. Schematic diagrams of the Erichsen test were shown in Figure 3. Samples for Erichsen test were machined into dimensions of 60 mm in length, 60 mm in width, and 1 mm in thickness.

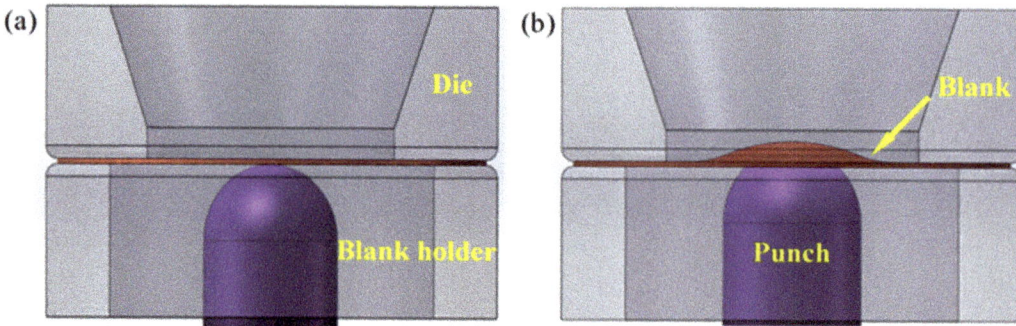

Figure 3. Schematic diagrams of the Erichsen test setup and punch position are shown (**a**) before and (**b**) after the test.

3. Results and Discussion

3.1. Microstructure of the As-Cast Alloy

The XRD patterns of the as-cast pure Mg and Mg-3Y alloys are shown in Figure 4. Both samples contain primary α-Mg phase. The additional diffraction peaks corresponding to the β-$Mg_{24}Y_5$ phases can be indexed in the Mg-3Y alloy.

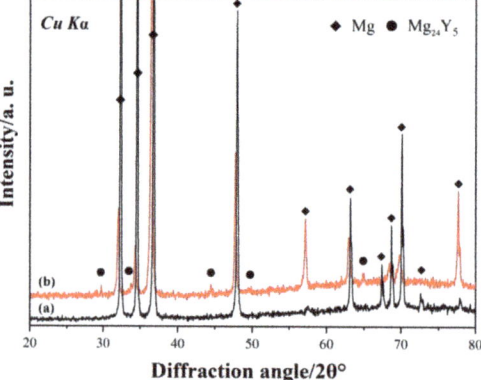

Figure 4. XRD patterns of the as-cast (**a**) pure Mg and (**b**) Mg-3Y alloys.

The macrostructures of as-cast pure Mg and Mg-3Y alloys are presented in Figure 5. The grain morphology changes from columnar grain to equiaxed grain and the grain size is clearly refined due to the addition of yttrium. The β-$Mg_{24}Y_5$ phases with a cubic crystal structure (lattice constant a = 11.257 Å) cannot act as the heterogeneous nuclei of α-Mg matrix [35]. However, the solute yttrium distributed on the solid–liquid interface could drag the dendrites growth during the growth of α-Mg grains, and thus refine the grains [36]. The growth restriction factor (GRF) could also be employed to describe the refining ability of solute elements in magnesium alloys. The GRF value of yttrium in magnesium is calculated to 1.7, which indicates that yttrium would have a certain refining ability [37].

The typical as-cast microstructure of the Mg-3Y alloy is shown in Figure 6. In Figure 6a,c, the Mg-3Y alloy is composed of α-Mg phase and numerous eutectic β-$Mg_{24}Y_5$ phases, which are discretely distributed between the dendrite arms. Most of the β-$Mg_{24}Y_5$ phases are bulk-shaped phases with different sizes, as shown in Figure 6b,d. The EDS result further confirms that the second phases existing in Mg-3Y alloy are β-$Mg_{24}Y_5$ phases, as shown in Figure 6d.

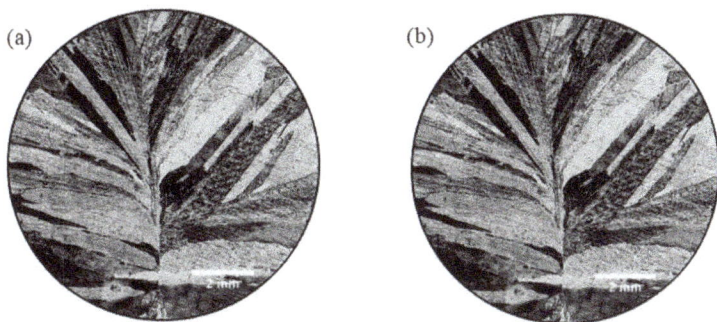

Figure 5. Macrostructures of the as-cast (**a**) pure Mg and (**b**) Mg-3Y alloys.

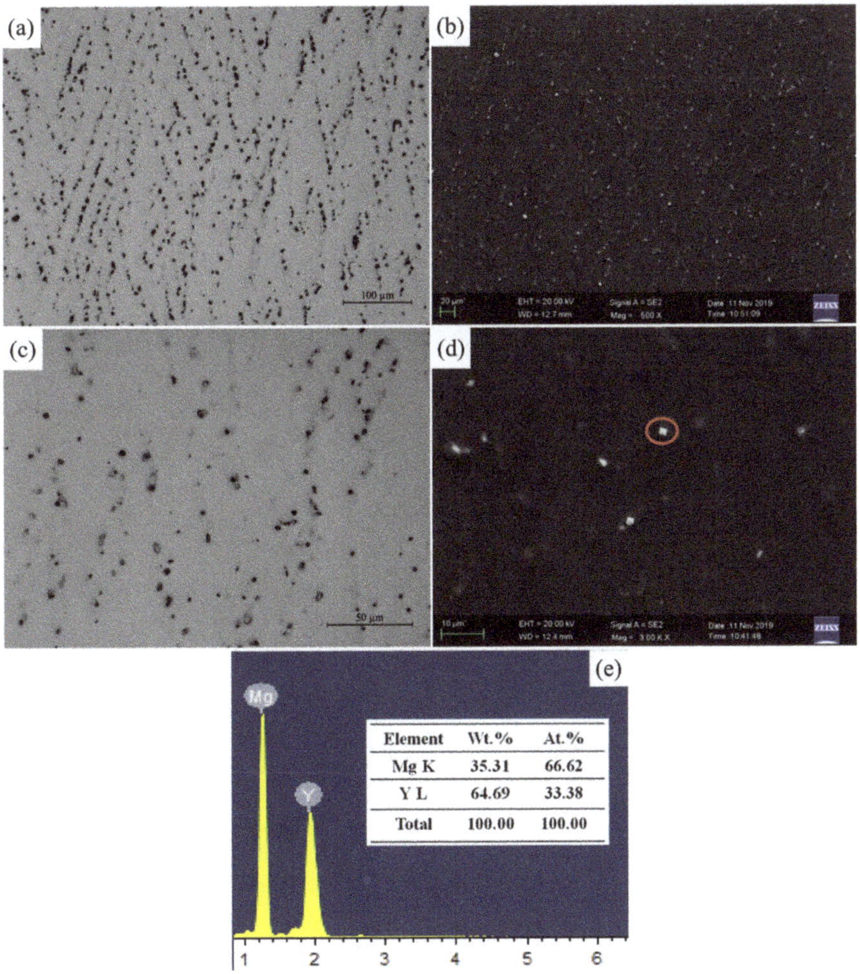

Figure 6. Optical micrographs of (**a**) low magnification view and (**b**) high magnification view; SEM micrographs of (**c**) low magnification view and (**d**) high magnification view; (**e**) the EDS results corresponded to the secondary phase marked by the red circle of as-cast Mg-3Y alloy.

The typical bulk-shaped β-Mg$_{24}$Y$_5$ phases with different sizes, from nanometer to micrometer, are presented in Figure 7. The size distribution of phases in nano-scale (Figure 7a,b) primarily varies from 225 to 675 nm (Figure 7e), accounting for about 58% of the total. In addition, the size distribution in micro-scale (Figure 7c,d) mainly changes from 1.12 to 2.48 μm (Figure 7e), accounting for about 34%. Although the size of the β-Mg$_{24}$Y$_5$ phases is fine, there is no clear aggregation. The morphological characteristic of β-Mg$_{24}$Y$_5$ phases with micro-nano size coexistence in matrix is formed finally.

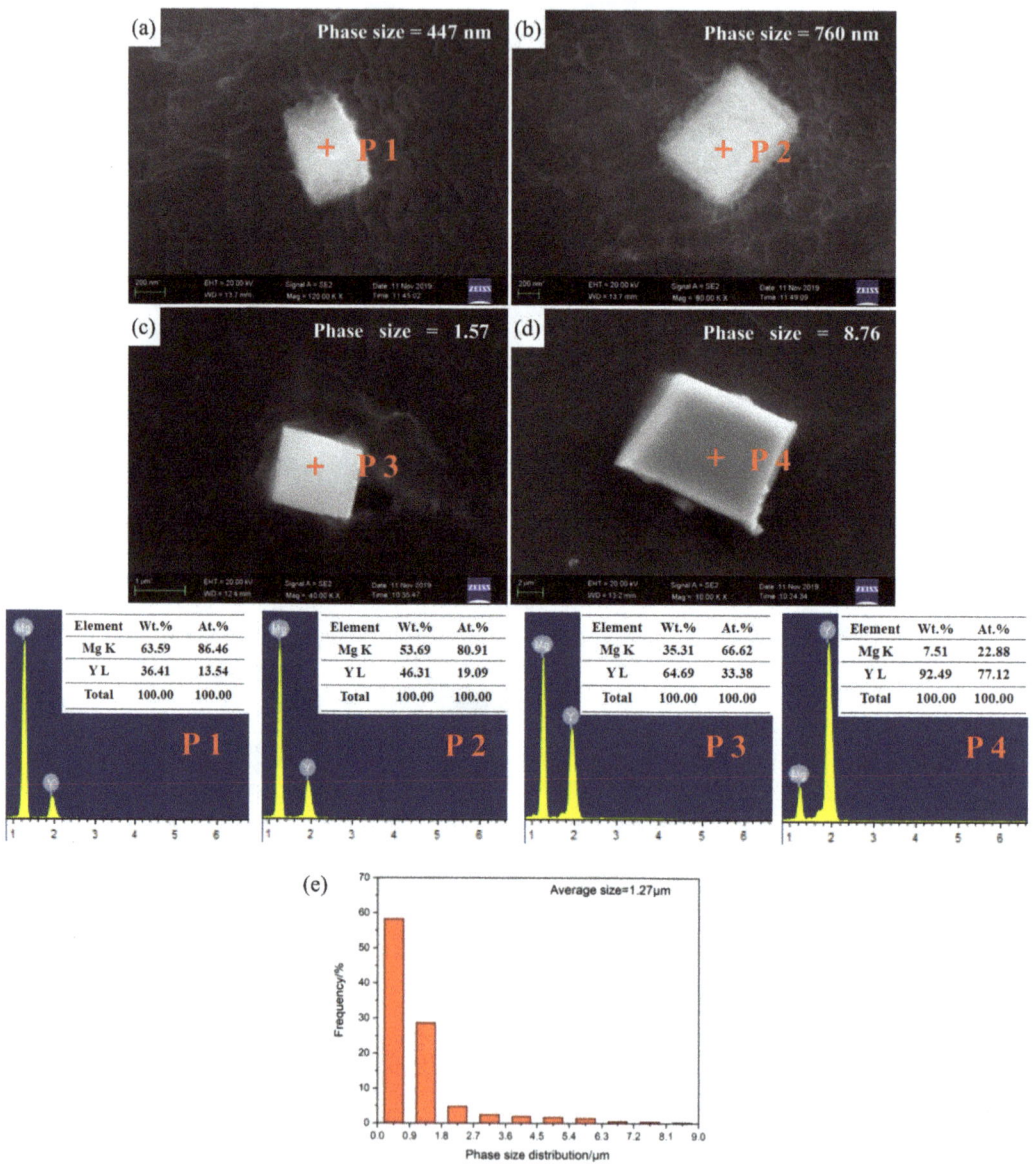

Figure 7. SEM micrographs and EDS results of the secondary phases with different sizes of (**a**,**b**) nano-scale, (**c**,**d**) micro-scale, and corresponding phase size distribution histograms (**e**) of as-cast Mg-3Y alloy.

Apart from micro-nano size coexistence characteristic, the Y-segregation layer is also observed at both nano- and micro-scales, as shown in Figure 8a,b. The β-$Mg_{24}Y_5$ phase core is surrounded by the semi-transparent light grey layer. It is speculated that the semi-transparent light grey layer might be an Y-segregation layer, which could provide further growth and development of β-$Mg_{24}Y_5$ phases. When the concentration of Y-segregation layer reaches a certain value, the β-$Mg_{24}Y_5$ phase core would consume the Y-segregation layer to further grow and develop. The β-$Mg_{24}Y_5$ phases with relatively larger size (about ≥2 μm) have no clear Y-segregation layer in Figure 7, which indicates the well development of β-$Mg_{24}Y_5$ phases. Therefore, the micro-nano size coexistence of β-$Mg_{24}Y_5$ phases should be relevant to the solidification segregation behavior of Y [38]. The segregation of Y during the solidification process results in the formation of Y-rich region (micro-phase) and Y-depleted region (nano-phase).

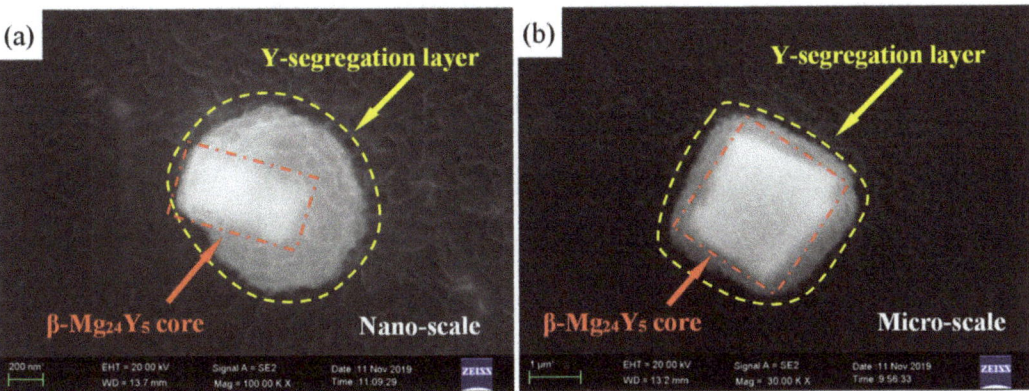

Figure 8. SEM micrographs of the secondary phases with Y-segregation layer of (**a**) nano-scale and (**b**) micro-scale of as-cast Mg-3Y alloy.

The nano- and micro-phase were further explored by TEM analysis. Figure 9a,b presents the bright-field TEM micrograph of nano phase-I and typical selected area electron diffraction (SAED) pattern of phase-I, respectively. It demonstrates that nano phase-I displays a rectangular shape, which is in good agreement with the morphology of β-$Mg_{24}Y_5$ phase, as shown in Figure 7a,b. In addition, the nano phase-I is a single crystal. The typical SAED pattern further ensures that phase-I corresponds to the β-$Mg_{24}Y_5$ phases. Notably, the interface between phase-I and Mg matrix is clear, which reveals that β-$Mg_{24}Y_5$ phases have good bonding with the matrix. Figure 9c,d shows the bright-field TEM micrograph of micro phase-I and local magnification of region A, respectively. Similarly, the micro phase-I also displays a rectangular shape and good bonding with the matrix. However, the micro phase-II consists of three parts (grain I–III) in addition to a single crystal. The β-$Mg_{24}Y_5$ phases are verified by the SAED pattern of grain-I and grain-II in Figure 9e,f. The high-resolution TEM (HR-TEM) observation was also performed at region B of grain-III, as can be seen in Figure 9g. The fast Fourier transform (FFT) and inverse FFT were conducted at the selected region C, correspondingly, as shown in Figure 9i,h. On the basis of the inverse FFT micrograph, the inter-planar spacing of grain-III is 0.301 nm, which is well consistent with the inter-planar spacing of ideal (321) plane for $Mg_{24}Y_5$ crystal. Consequently, the micro phase-II is polycrystalline with three grains rather than the clustering of nano-phases. The nano grain-III is covered by grain-I and grain-II. At the same time, the nano grain-II is covered by micro grain-I. This phenomenon may indicate that the micro-phase is developed from the nano-phase, which is well consistent with the solidification segregation behavior of Y, as mentioned above.

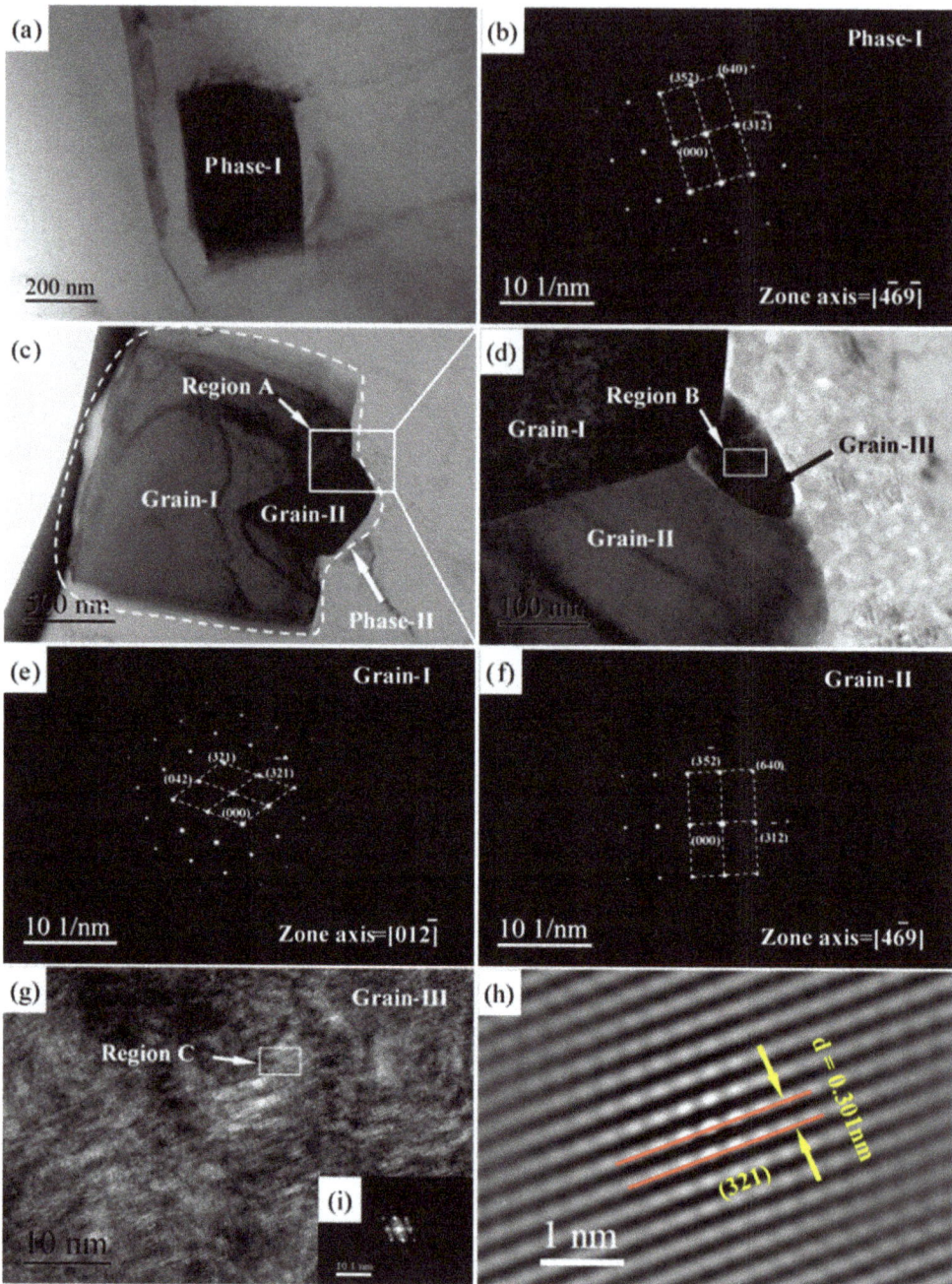

Figure 9. Bright field TEM micrographs of (**a**) phase-I, (**c**) phase-II, and (**d**) local magnification corresponding to region A; SAED patterns of (**b**) phase-I, (**e**) grain-I, and (**f**) grain-II; (**g**) HRTEM micrograph of region B in grain-III; (**i**) FFT pattern; and (**h**) inverse FFT micrograph of region C.

3.2. Microstructures and Micro-Texture Evolution of Alloy Sheets

The optical micrographs of Mg-R and Mg-RA sheets are shown in Figure 10. The microstructure of as-rolled samples is heterogeneous with a distribution in grain size from several μm to hundreds of μm. Apart from some dynamic recrystallized (DRXed) grains induced via grain boundary, many twins exist in coarse deformed grains. For as-annealed samples, the heterogeneity of grain size is faded along with grain coarsening.

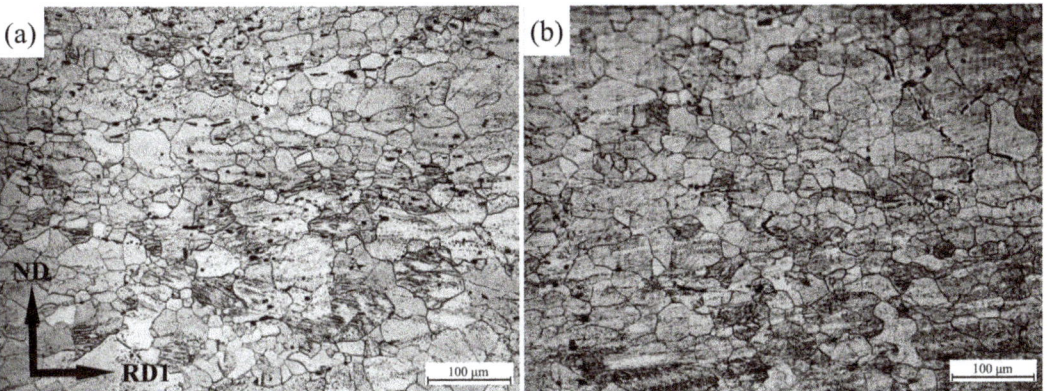

Figure 10. Optical micrographs of (**a**) as-rolled and (**b**) as-annealed pure Mg sheets.

Figure 11a–d shows the microstructure of Mg-3Y-R and Mg-3Y-RA sheets. The microstructure of Mg-3Y-R sheet contains deformed grains dotted with fine and dispersed β-$Mg_{24}Y_5$ phases. The β-$Mg_{24}Y_5$ phases exist both at grain boundaries and in the interior of grains. Micron sized β-$Mg_{24}Y_5$ phases have no clear change and preserve the original morphological characteristic under as-cast condition (Figure 11e). Moreover, no apparent dynamic recrystallization (DRX) is found. In fact, the large second-phase particles (>1 μm) are the ideal sites for the development of DRX nuclei by particle stimulated nucleation (PSN) [39,40]. However, the PSN effect is absent around the micron-sized β-$Mg_{24}Y_5$ phases (~2.74 μm), as shown in Figure 11e. The fragmentation of β-$Mg_{24}Y_5$ phases is observed in Figure 11f, which indicates that nano-sized phases with weak bonding have been broken to some extent during rolling. The directional distribution of dispersed β-$Mg_{24}Y_5$ phases emerges. After annealing, the deformed grains change to equiaxed grains with uniform size for Mg-3Y-RA sheet due to static recrystallization (SRX). The distribution characteristic of β-$Mg_{24}Y_5$ phases has no significant difference from Mg-3Y-R sheets. In comparison with the Mg-R and Mg-RA sheets (see Figure 10), the Mg-3Y-R and Mg-3Y-RA sheets exhibit considerably more homogenous microstructures favoring further deformation.

The twinning is depressed with increasing deformation temperature [41] and extensive wrinkled structures occur in Mg-3Y-R sheet rolled at a quite high temperature of 773 K, as shown in Figure 12a,c. The wrinkled structures present lath structures with different contrasts in TEM. No diffraction spots for twins are detected and the interface of these lath structures is not as straight as twin boundaries, as indicated by Figure 12e. Therefore, these lath structures should be deformation bands, but twins. The deformation bands are almost parallel and the top is thin and sharp. Yang X Y et al. [5,42] believed that the deformation bands are kink bands, which correspond to a low angle dislocation interface. The kink bands are formed by bending the slip surface when the primary slip is blocked. The formation of kink bands is closely related to crystal orientation, deformation temperature, and deformation degree. Similar results have been reported by Zhou B et al. [43] and Matsumoto T et al. [44]. After annealing, the density number of kink bands decreases remarkably (Figure 12b,d) and the residual kink bands are discontinuous (Figure 12f).

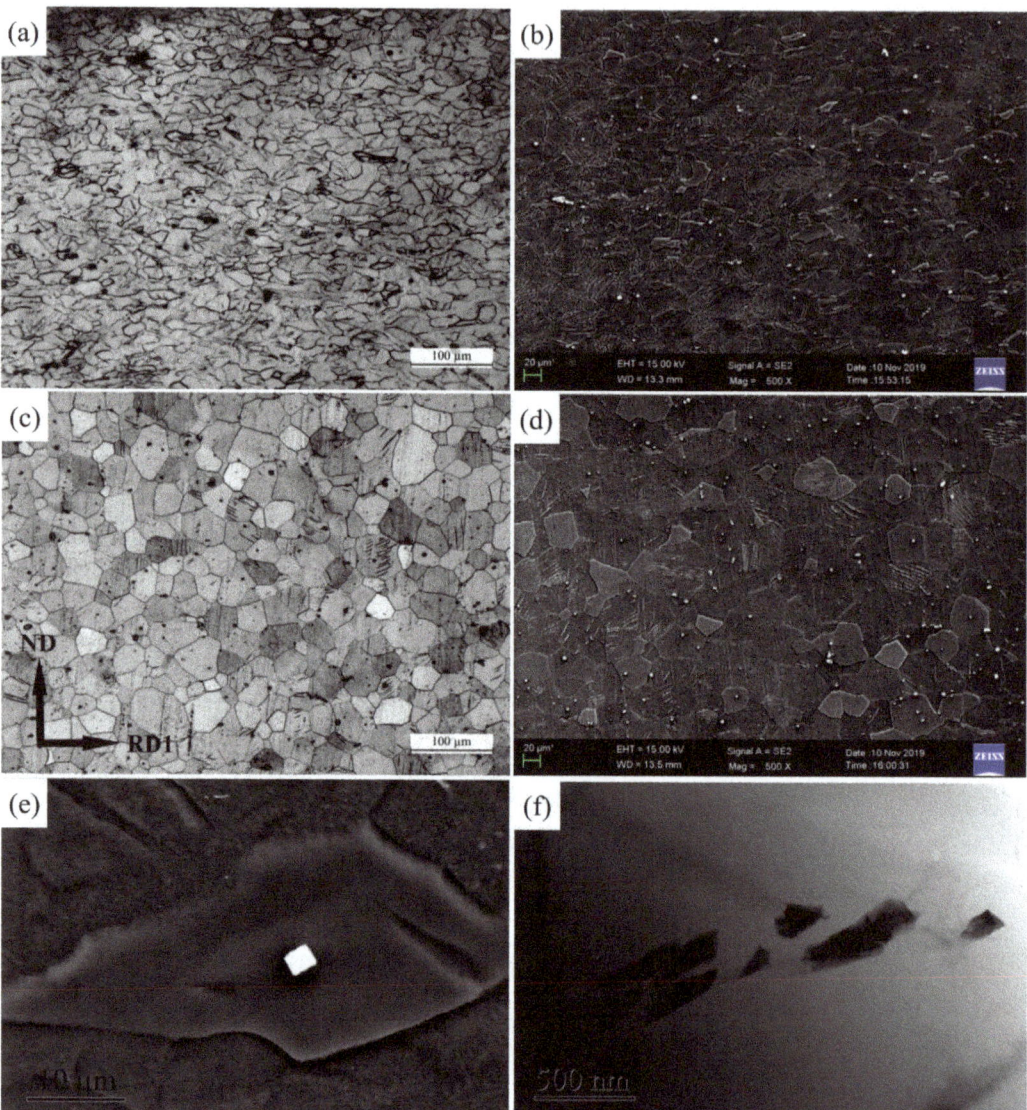

Figure 11. Optical micrographs of (**a**) as-rolled, (**c**) as-annealed, and corresponding SEM micrographs of (**b**) as-rolled and (**d**) as-annealed Mg-3Y sheets, the secondary phases; (**e**) SEM micrograph and (**f**) TEM micrograph of Mg-3Y sheets.

EBSD measurements were conducted to further analyze the evolution of the microstructure and micro-texture on the RD1-ND plane of the alloy sheets. Figure 13a–d presents the EBSD inverse pole figure (IPF) maps and corresponding grain size distribution of Mg-3Y sheets. Different colors represent different orientations of the grains. The grains with similar color indicate that the misorientation angles between these grains are not significant in Mg-3Y-R sheets. In addition, the average grain size is about 9.76 μm. After annealing, the grains with multicolor intuitively show the weakening of the preferred orientation of grains. Moreover, the average grain size increases to about 13.8 μm. Figure 13e,f shows (0001), (11-20), and (10-10) pole figures of Mg-3Y sheets. The micro-texture characteristic of Mg-3Y-R sheet is the (0002) basal texture with a split of basal pole along the RD2 direction. The (0002) basal texture is weakened for Mg-3Y-RA sheet after annealing. First, the basal pole splits into multiple peaks and the orientation distribution becomes broader. The (11-20) and (10-10) planes have a random orientation distribution. Second, the maximum pole intensity decreases from 7.8 to 3.5 multiples of random orientation distribution (MRD).

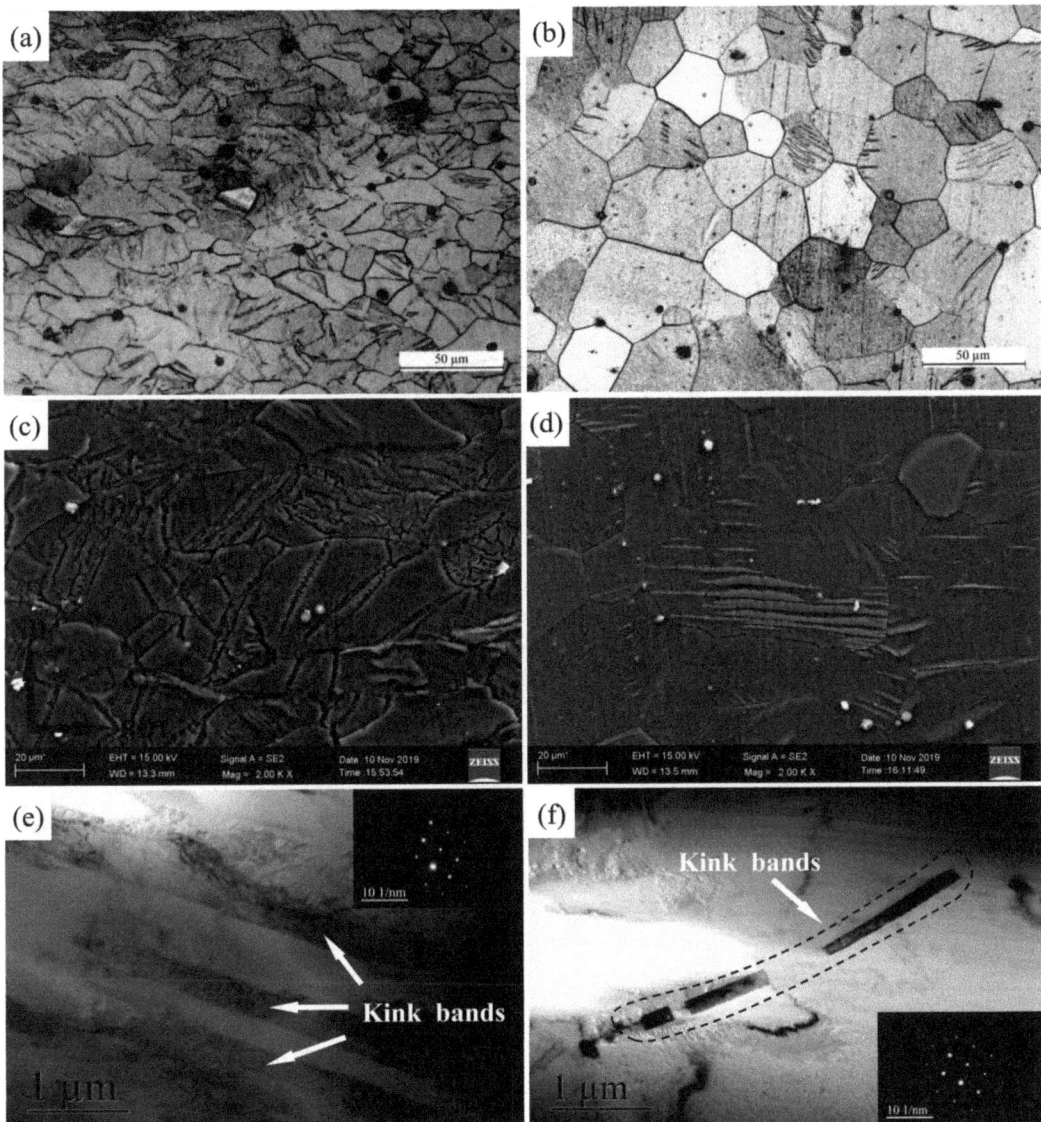

Figure 12. Optical micrographs of (**a**) as-rolled, (**b**) as-annealed, and corresponding SEM micrographs of (**c**) as-rolled and (**d**) as-annealed Mg-3Y sheets; TEM micrographs of kink bands of (**e**) as-rolled and (**f**) as-annealed Mg-3Y sheets.

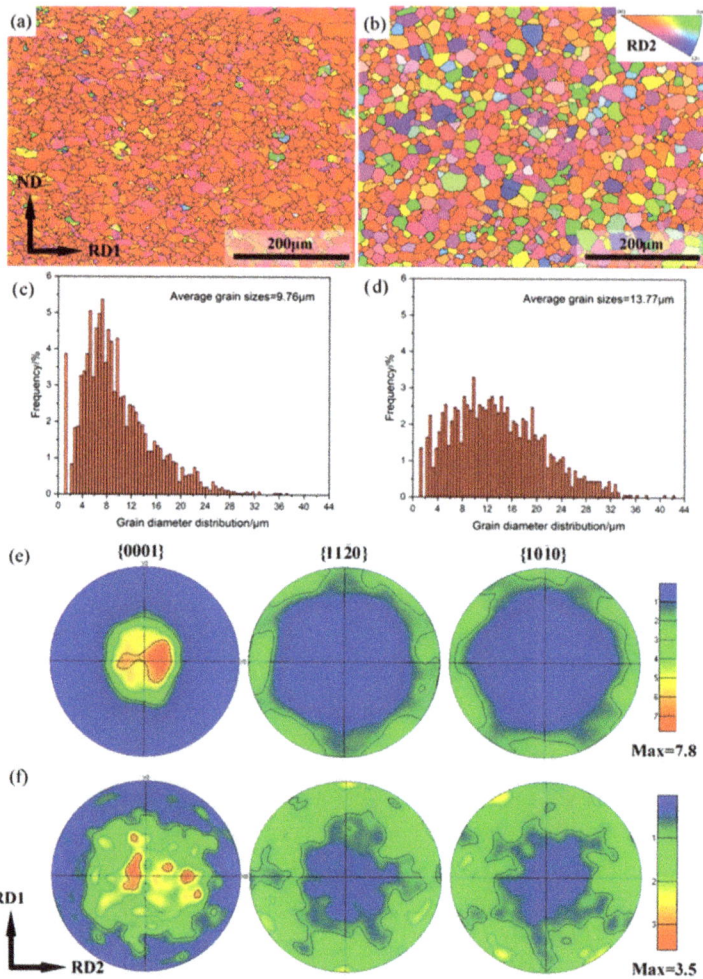

Figure 13. EBSD IPF maps of (**a**) as-rolled and (**b**) as-annealed; grain size distributions of (**c**) as-rolled and (**d**) as-annealed; (0001), (11-20), and (10-10) pole figures of (**e**) as-rolled and (**f**) as-annealed Mg-3Y sheets. Observation along RD2 was applied to IPF triangle.

Grain boundary maps and misorientation angle distributions of Mg-3Y sheets are shown in Figure 14a–d. The high-angle grain boundaries (HAGBs > 15°) and low-angle grain boundaries (LAGBs < 15°) are characterized by black and green lines, respectively. Combined with the misorientation angle distributions, it can be seen that the proportion of LAGBs (78.72%) is considerably higher than HAGBs (21.28%) in Mg-3Y-R sheet. Indeed, the number fraction of boundaries with low misorientation angles (<10°), which are a result of lattice distortion and subgrain formation within grains caused by dislocation slip, is 75.12% and dominates the major proportion. Intersection of {10-12} extension twins can lead to the formation of twin boundaries with a low misorientation angle of ~7.4° [45]. However, a comparison of the twin boundary map (described below in Figure 15a) with the low-angle boundary map reveals that Mg-3Y-R sheet does not contain low-angle boundaries formed by this twin intersection. In contrast, the proportion of HAGBs (88.20%) is considerably higher than LAGBs (11.80%) in Mg-3Y-RA sheet. Based on [46], the proportion of HAGBs is related to the amount of recrystallized and deformed grains in the microstructure. A large amount of recrystallized grains would result in high proportion of HAGBs, meanwhile a large amount of deformed grains would result in high proportion of LAGBs. The recrystallization fraction of Mg-3Y-RA sheet is almost 100%, but the recrystallization fraction is only 4.79% for Mg-3Y-R sheet (Figure 14e–h). Relative complete SRX process produces a greater proportion of HAGBs. Furthermore, this results in more random grain orientation distribution and clear weakening of the (0002) texture.

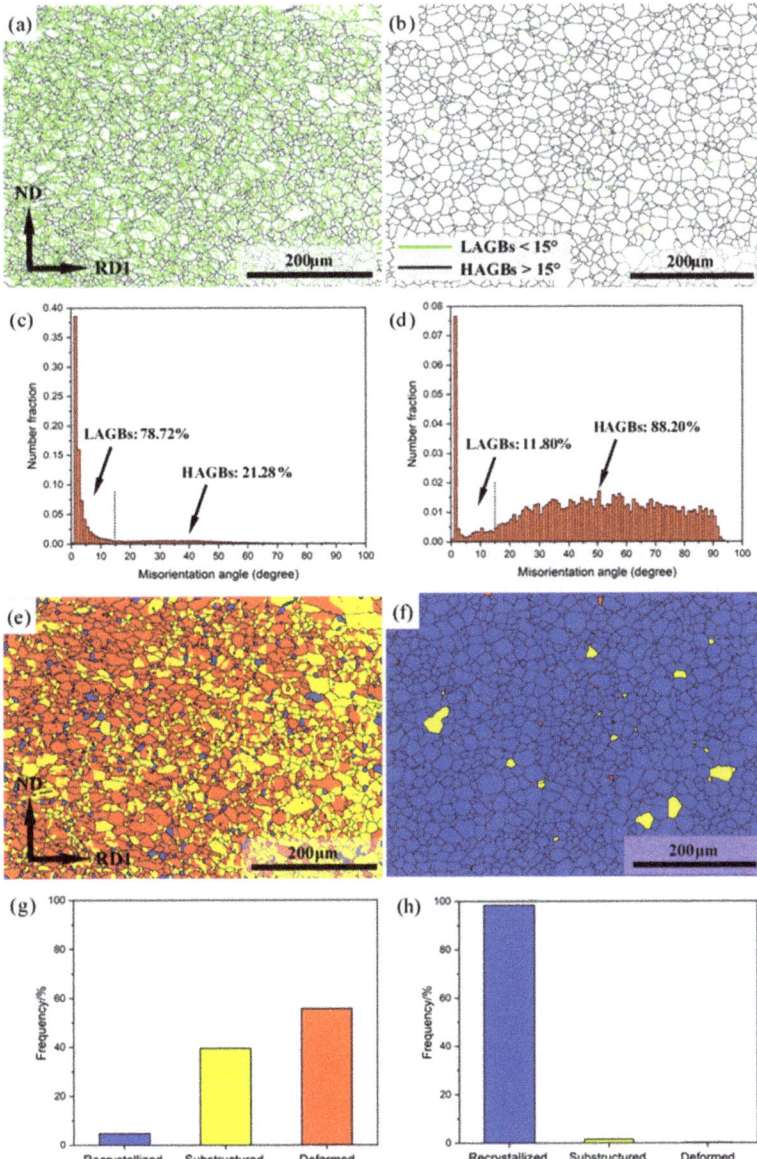

Figure 14. Grain boundary maps of (**a**) as-rolled and (**b**) as-annealed; misorientation angle distributions of (**c**) as-rolled and (**d**) as-annealed; structure component distribution maps of (**e**) as-rolled and (**f**) as-annealed; and structure fraction distributions of (**g**) as-rolled and (**h**) as-annealed Mg-3Y sheets.

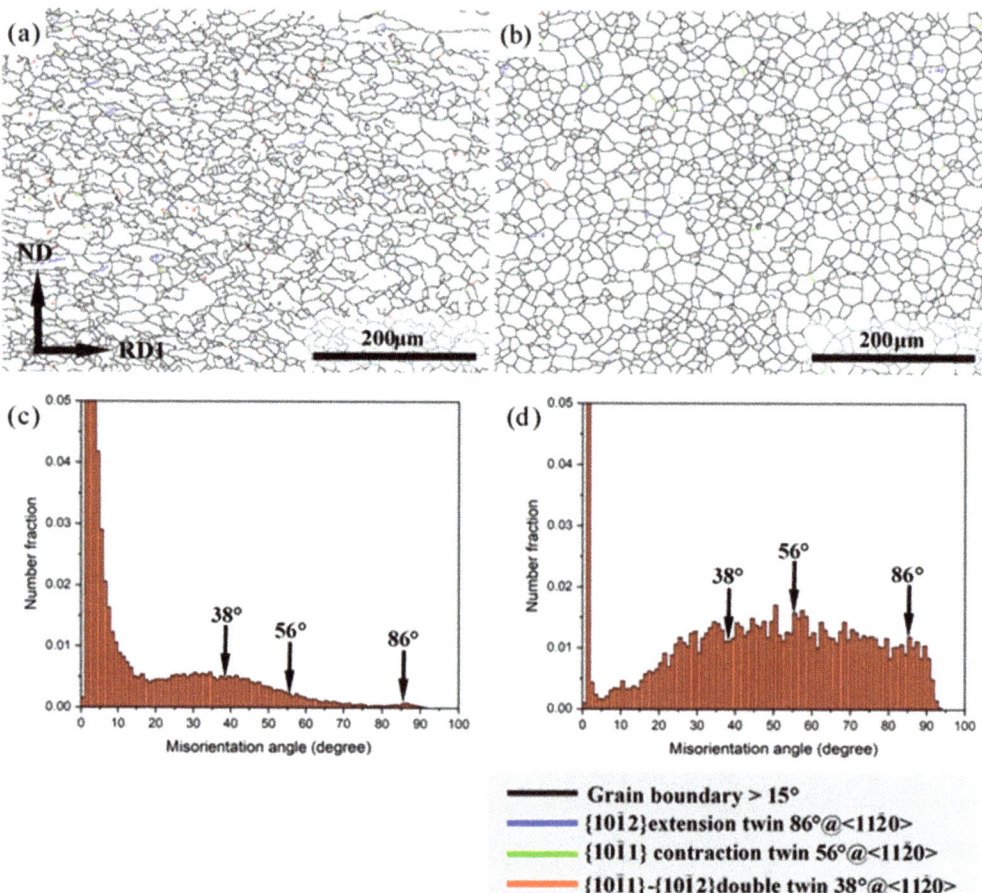

Figure 15. Misorientation angle maps showing various twin boundaries of (**a**) as-rolled and (**b**) as-annealed; misorientation angle distributions of (**c**) as-rolled and (**d**) as-annealed Mg-3Y sheets.

Misorientation angle maps depicting various twin boundaries and corresponding misorientation angle distributions of Mg-3Y sheets are shown in Figure 15. The main twin types present in Mg-3Y sheets are {10-12} extension twin, {10-11} contraction twin, and {10-11}-{10-12} double twin. Despite the appearance of the different types of twins in both Mg-3Y-R and Mg-3Y-RA sheets, the fraction of twins is very small. The proportions of {10-12} extension twin, {10-11} contraction twin, and {10-11}-{10-12} double twin are 0.08%, 0.18%, and 0.47% for Mg-3Y-R sheet, respectively. The total proportion of twins is about 0.73%, less than 1.0%. As for Mg-3Y-RA sheet, the proportions of {10-12} extension twin, {10-11} contraction twin, and {10-11}-{10-12} double twin are 1.01%, 1.49%, and 1.11%, respectively. Although the total proportion (about 3.61%) of twins has some increase after annealing, the proportion is still very small. The increase in twins proportions may be attributed to the following two possible reasons. The reduction of total grain boundaries quantity results from the growth of grains and the preservation of twins during annealing. In addition, the preservation of twins indicates that the twins cannot act as the favorable nucleation site for SRX during subsequent short-term annealing. Combined with the analysis about the change of morphologies and number density of kink bands, the kink bands should be the favorable nucleation site for SRX except for grain boundaries. As mentioned above, few twins prove that the deformation bands should be kink bands, but twins again.

The kernel average misorientation (KAM) maps and corresponding local misorientation distributions of Mg-3Y sheets are shown in Figure 16. The KAM map is constructed based on the average misorientation between a measuring point and all its neighbors. Consequently, the local misorientation and strain energy can be clearly reflected in the KAM map [47,48]. In the KAM maps, the KAM values indicate local misorientation, which we interpret in terms of density of geometrically necessary dislocations to provide the local strain value (high KAM value = high misorientation = high strain value). The high strain energy zone could be clearly observed by the red and orange colors. The relatively uniform distribution of similar colors indicates that the deformation of Mg-3Y-R sheet is comparatively homogenous except for a few high strain energy zones. The average KAM value of Mg-3Y-R sheet is found to be higher and drops from 1.57 to 0.38 remarkably after subsequent short-term annealing. Therefore, most of the areas nearly exhibit a strain-free state, indicating that SRX is a process of softening the work hardening effect [49,50]. The release of residual stress is beneficial to the improvement of ductility and stretch formability of Mg-3Y sheets.

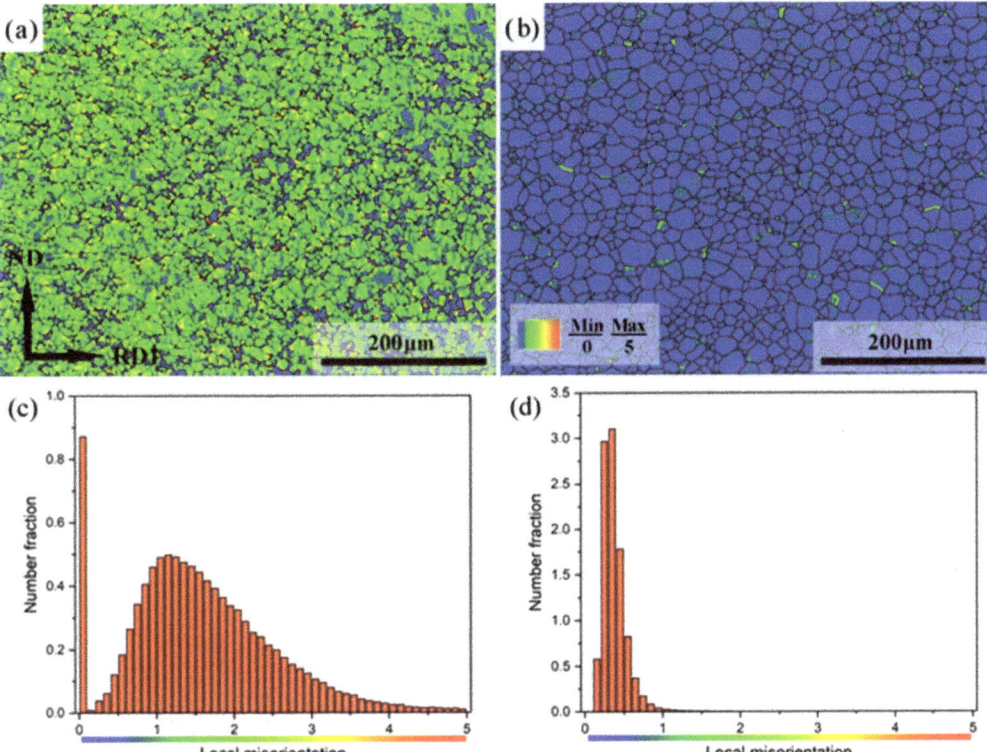

Figure 16. KAM maps of (**a**) as-rolled, (**b**) as-annealed, and corresponding local misorientation distributions of (**c**) as-rolled and (**d**) as-annealed Mg-3Y sheets.

To fully understand the effect of annealing on the ductility of Mg-3Y sheets, the Schmid factor (SF) distributions of the (0001)<11-20> basal <a> slip, (1-100)<11-20> prismatic <a> slip, (1-101)<11-20> pyramidal <a> slip, and (11-22)<11-2-3> pyramidal <c+a> slip is investigated by EBSD analysis. Figure 17 shows the (0001)<11-20> basal <a> slip SF maps and corresponding SF distributions of Mg-3Y sheets. The blue grains usually have their basal planes parallel to the tensile direction, i.e., along the RD2 direction. The blue grains have a relatively lower basal <a> slip SF and is unfavorable for basal <a> slip. However, the red grains usually have their basal planes inclined by about 45° to the tensile planes. Therefore, the red grains have a higher basal <a> slip SF, which is favorable for basal <a> slip. Compared with Mg-3Y-R sheet, a large number of red grains are observed in Mg-3Y-RA sheet, which is correlated with the weak texture resulting from SRX during annealing. As can be seen from the fraction distribution of SF for basal <a> slip of both sheets before and after annealing, the number fraction of grains gradually decreases with the increasing SF values (with regard to SF values variation range of 0.3–0.5) in Mg-3Y-R sheet. The number fraction of grains with the highest SF values (> 0.4) [51] corresponds to the lowest number fraction of 7.17%. In contrast, there is a gradual increase in SF distribution for Mg-3Y-RA sheet. The number fraction of grains with a high SF value (> 0.4) is greater than 38.24%. Correspondingly, the average SF for basal <a> slip increases from 0.21 to 0.31 after annealing. Therefore, the activation of (0001)<11-20> basal <a> slip is further enhanced in Mg-3Y-RA sheet.

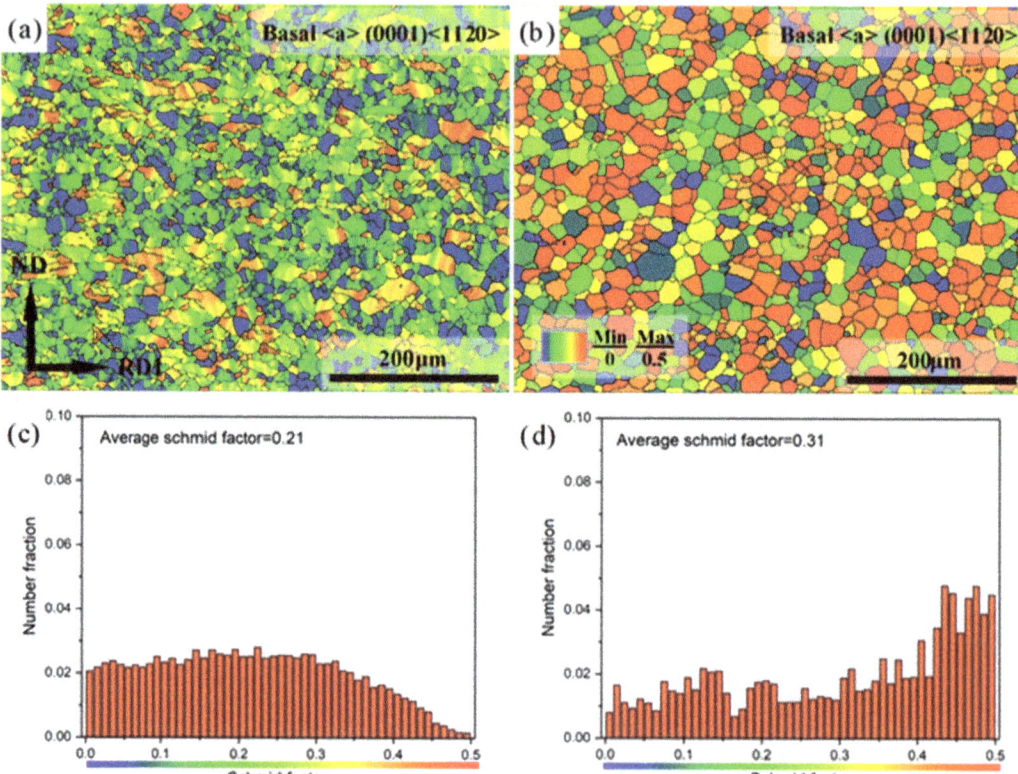

Figure 17. Basal <a> slip SF maps of (**a**) as-rolled, (**b**) as-annealed, and corresponding SF distributions of (**c**) as-rolled and (**d**) as-annealed Mg-3Y sheets.

The (1-100)<11-20> prismatic <a> slip SF maps and corresponding SF distributions of Mg-3Y sheets are shown in Figure 18. As mentioned above, the prismatic planes of blue grains are usually parallel to the tensile direction and have relatively lower prismatic <a> slip SF, which is unfavorable for prismatic <a> slip. The prismatic planes of red grains are usually inclined by about 45° to the tensile planes and have a higher prismatic <a> slip SF, which is favorable for prismatic <a> slip. The colors of all grains are approaching the red color, which indicates that prismatic <a> slip SF values are high for Mg-3Y-R sheet. The number fraction of grains with high prismatic <a> slip SF values (>0.4) is about 84.13% and almost all the SF values of grains are more than 0.3. Nevertheless, the number fraction of grains with high prismatic <a> slip SF values decreases to 46.15% for Mg-3Y-RA sheet. In addition, some grains show unfavorable orientation for prismatic <a> slip. Nevertheless, the average SF for prismatic <a> slip decreases from 0.44 to 0.35 after annealing. Moreover, the activation of (1-100)<11-20> prismatic <a> slip (smaller value of average SF: 0.35) is still higher than (0001)<11-20> basal <a> slip (larger value of average SF: 0.31) for both Mg-3Y sheets.

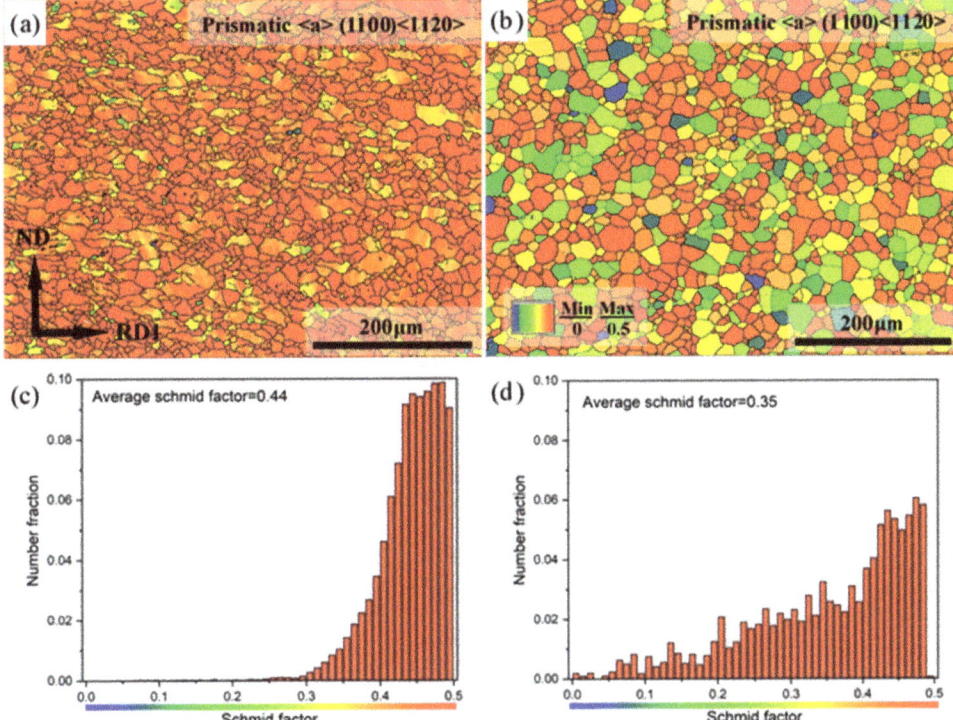

Figure 18. Prismatic <a> slip SF maps of (**a**) as-rolled, (**b**) as-annealed, and corresponding SF distributions of (**c**) as-rolled and (**d**) as-annealed Mg-3Y sheets.

Figure 19 shows the pyramidal slip (including pyramidal <a> slip and pyramidal <c+a> slip) SF maps and corresponding SF distributions of Mg-3Y sheets. Similarly, blue grains with relatively lower pyramidal slip SF are unfavorable for pyramidal slip. The red grains with a higher pyramidal slip SF are favorable for pyramidal slip. The (1-101)<11-20> pyramidal <a> slip SF maps and corresponding SF distributions are displayed in Figure 19a–d. The number fractions of grains with high pyramidal <a> slip SF values (>0.4) are about 69.05% and 68.33% for Mg-3Y-R and Mg-3Y-RA sheets, respectively. Nearly all the SF values of grains are more than 0.3 for both Mg-3Y-R and Mg-3Y-RA sheets. The average SF values for pyramidal <a> slip have no clear change and maintain the same level before and after annealing. Therefore, annealing has little effect on the activation of (1-101)<11-20> pyramidal <a> slip. The (11-22)<11-2-3> pyramidal <c+a> slip SF maps and corresponding SF distributions are displayed in Figure 19e–f. The number fractions of grains with high pyramidal <c+a> slip SF values (>0.4) are about 56.75% for Mg-3Y-R sheets. After annealing, the number fraction of grains with high pyramidal <c+a> slip SF values drops to 29.48% for Mg-3Y-RA sheet.

Accordingly, the average SF for pyramidal <c+a> slip decreases from 0.40 to 0.33 after annealing. Similar to (1-100)<11-20> prismatic <a> slip, the activation of (11-22)<11-2-3> pyramidal <c+a> slip (smaller value of average SF: 0.33) is higher than (0001)<11-20> basal <a> slip (larger value of average SF: 0.31) for both Mg-3Y sheets.

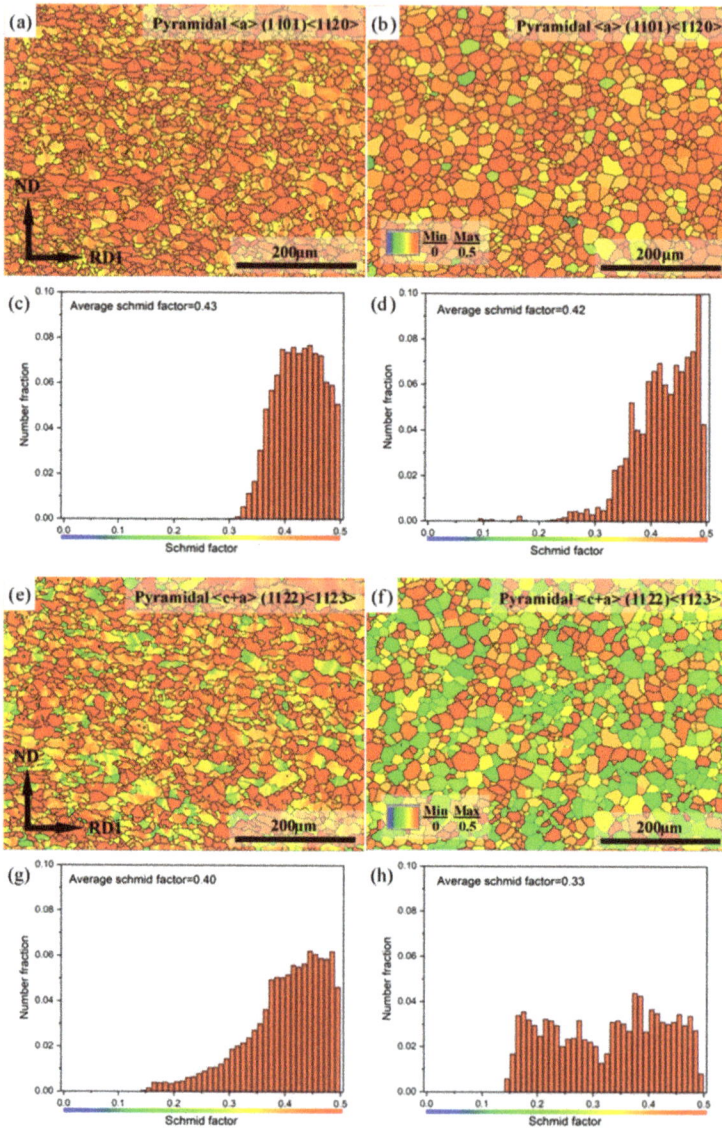

Figure 19. Pyramidal <a> slip SF maps of (**a**) as-rolled, (**b**) as-annealed, and corresponding SF distributions of (**c**) as-rolled and (**d**) as-annealed; pyramidal <c+a> slip SF maps of (**e**) as-rolled, (**f**) as-annealed, and corresponding SF distributions of (**g**) as-rolled and (**h**) as-annealed Mg-3Y sheets.

Combined with the analysis of SF above, it can be concluded that the addition of Y into Mg promotes the activity of non-basal slip [52,53]. Both average SF values of prismatic <a> slip, pyramidal <a> slip, and pyramidal <c+a> slip are greater than 0.30 for Mg-3Y-R and Mg-3Y-RA sheets, which show a high activation (Table 2). The Mg-3Y-RA sheet has a crystallographic orientation, which is more favorable for basal <a> slip, but a little unfavorable for prismatic <a> slip and pyramidal <c+a> slip after SRX. Therefore, the average SF values of basal <a> slip increase, but decrease for prismatic <a> slip and pyramidal <c+a> slip. The enhanced activity of basal <a> slip could be beneficial for the improvement of ductility and stretch formability at room temperature for Mg-3Y-RA sheet.

Table 2. Average Schmid factor (SF) of slip systems in Mg-3Y sheets.

Sheets	Average Schmid Factor (SF)			
	Basal <a> (0001)<11-20>	Prismatic <a> (1-100)<11-20>	Pyramidal <a> (1-101)<11-20>	Pyramidal <c+a> (11-22)<11-2-3>
Mg-3Y-R	0.21	0.44	0.43	0.40
Mg-3Y-RA	0.31	0.35	0.42	0.33

3.3. Macro-Texture

The macro-texture of the sheets by means of the (0002) pole figures are displayed in Figure 20. As expected, a typical strong basal texture is formed in Mg-R sheet, where the orientation distribution of most basal poles is parallel to the normal direction (Figure 20a). After annealing, the texture intensity increases from 10.4 to 13.4 MRD, owing to the coarsening of grain in Mg-RA sheet, as shown in Figure 20b.

In contrast, the qualitative character of the Mg-3Y-R sheet texture is clearly distinct from typical basal texture. The texture has a clearly broader distribution of basal poles, as compared with its distribution in the Mg-R sheet. It was reported that the basal pole for many Mg alloys containing RE elements would be simple to spread from ND toward transverse direction (TD), resulting in an ellipse-shape orientation distribution of the (0002) basal texture [11,54]. However, in our cross-rolling case, the RD1 and RD2 in turn serve as the TD in a conventional unidirectional rolling. The basal pole would spread toward RD2 and RD1 by turning. Therefore, the (0002) basal texture of Mg-3Y-R sheet shows a subrotund orientation distribution (Figure 20c), which is beneficial to reduce the planar anisotropy between TD and RD [55]. Furthermore, the Mg-3Y-R sheet shows a split of the basal-pole intensity peak from ND toward RD2, which results in the formation of a double peak texture. This RD-splitting characteristic may be considered as the result of activation of pyramidal <c+a> slip [7,56]. Compared with Mg-R sheet, the maximum intensity of Mg-3Y-R sheet decreases from 10.4 to 5.2 MRD and the new texture components are formed. To reveal the change of texture and components, the orientation distribution function (ODF) sections of Mg-3Y-R sheet are presented in Figure 21. The major texture components of Mg-3Y alloy sheets are summarized in Table 3. There are three relatively stronger texture fibers with some dominant components in Mg-3Y-R sheet. The first consists of main texture component of A-{01-17}<-1-231> orientation grains, the second consists of main texture component of B-{0001}<-1-231> orientation grains, and the third consists of main texture component of C-{10-17}<-1-231> orientation grains. After annealing, a considerably weaker sheet texture is obtained in Mg-3Y-RA sheet compared with the Mg-3Y-R sheet. The orientation distribution of basal poles becomes more randomized and the (0002) basal maximum intensity further reduces to 3.1 MRD. The tendency of macro-texture intensity variety is the same as the micro-texture mentioned above. The double peak texture disappears and a new multiple-peak texture appears. High deformation temperature enhances the activities of non-basal slips and GBS, which may affect the SRX kinetics. Deformation at higher temperatures increases the orientation gradients near grain boundaries due to GBS and/or shearing connected with grain boundary serration [57]. In the previous work, some of the authors reported that a significant weakening of texture may be achieved by annealing a rolled AZ31 alloy with a deformation microstructure without an occurrence of DRX [23,24], which has also been revealed to be due to discontinuous SRX at pre-existing grain boundaries, i.e., the large orientation gradients and the high local dislocation densities near grain boundaries are likely to induce SRX at pre-existing grain boundaries [58]. Therefore, it can be expected that the textures of the sheet rolled at higher temperatures weaken more remarkably during annealing if plenty of pre-existing grain boundaries are contained. Consequently, strong SRX induced by plenty of pre-existing grain boundaries should be the reason for the change of texture character in Mg-3Y-RA sheet. In the same way, the ODFs of Mg-3Y-RA sheet are presented in Figure 22 to reveal the change of texture and components after annealing. The texture fibers and components vanish and numerous weakened discrete textures and components emerge. Not only the scatter is increased, but also the intensity is significantly decreased. There are seven relatively stronger discrete textures with different dominant components (D-J orientation grains) in Mg-3Y-RA sheet. As pointed by Suh B C et al. [59,60], this multiple-peak texture is beneficial in obtaining good formability since a large number of grains are favorably oriented for basal <a> slip to operate during stretch forming in comparison with the RD-split texture developed in the AXM100 alloy.

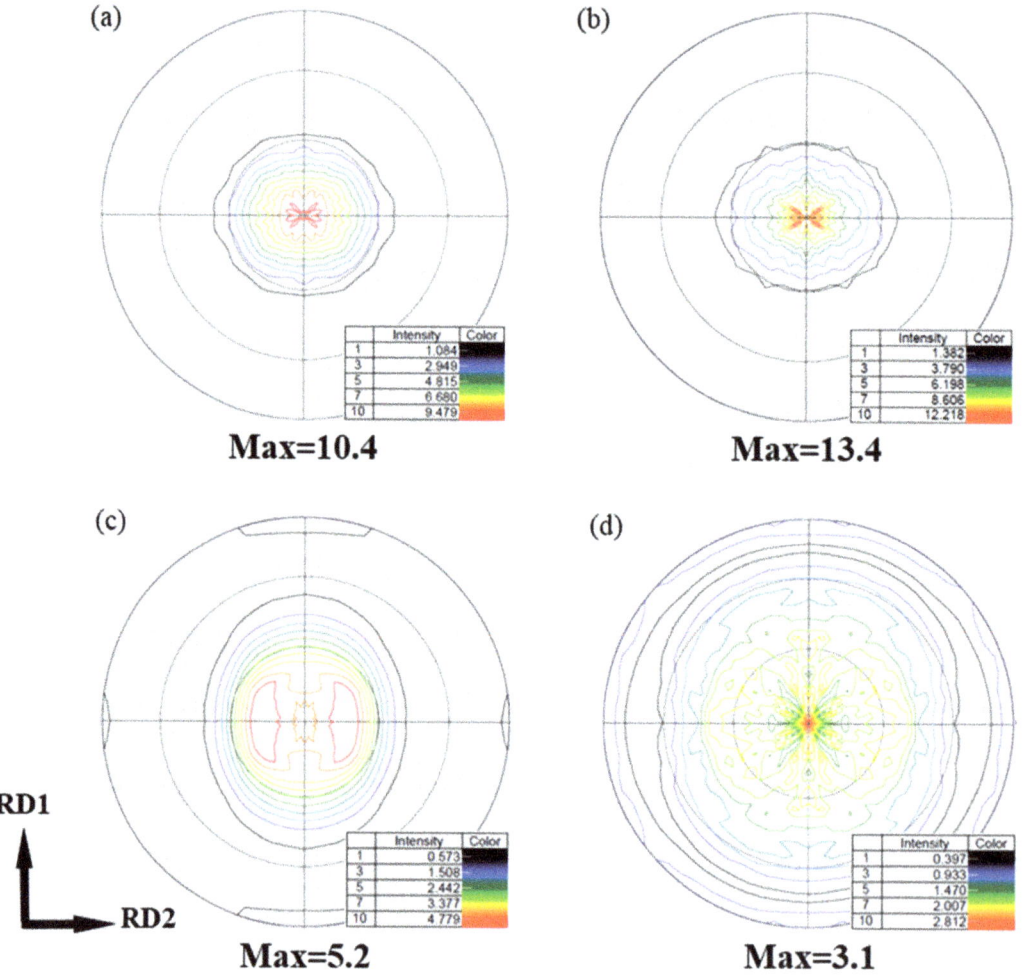

Figure 20. X-ray recalculated (0002) pole figure of (**a**) as-rolled and (**b**) as-annealed pure Mg sheets; (**c**) as-rolled and (**d**) as-annealed Mg-3Y sheets.

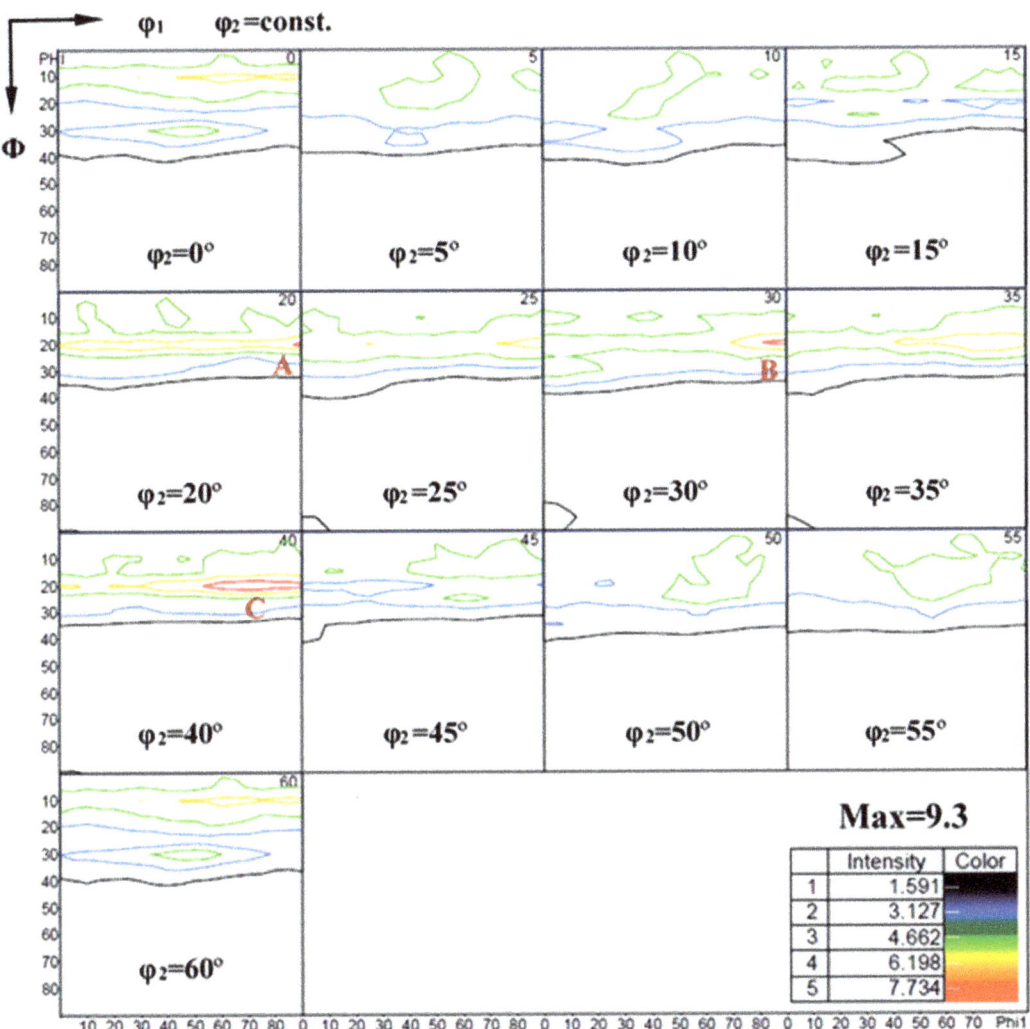

Figure 21. ODF sections of as-rolled Mg-3Y sheet (A, B and C are different texture component mark).

Table 3. Major texture components of Mg-3Y alloy sheets.

Mg-3Y-R Sheet		Mg-3Y-RA Sheet	
Symbol	Texture Components	Symbol	Texture Components
		D	{01-11}<2-1-10>
		E	{01-11}<8-7-16>
A	{01-17}<-1-231>	F	{01-13}<0-332>
B	{0001}<-1-231>	G	{11-26}<1-542>
C	{10-17}<-1-231>	H	{10-11}<1-210>
		I	{10-11}<0-111>
		J	{10-13}<-4-153>

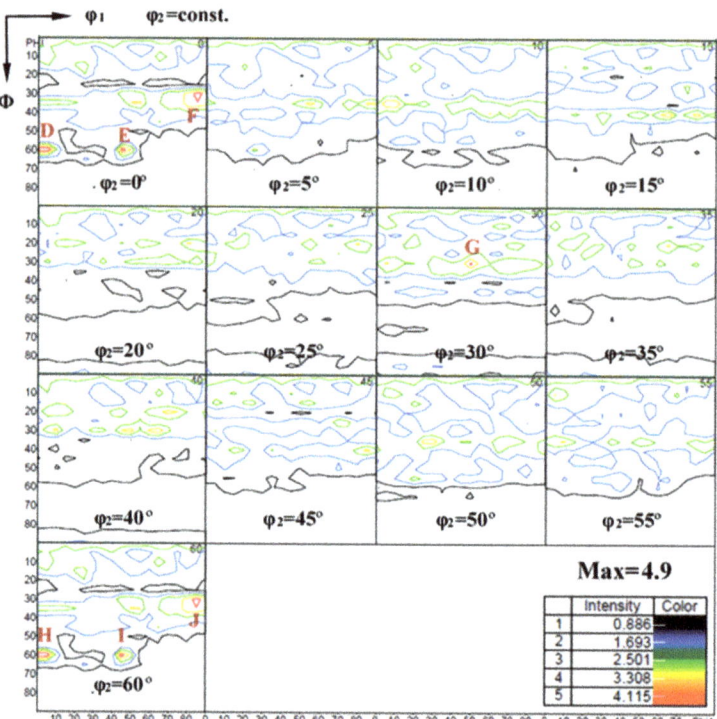

Figure 22. ODF sections of as-annealed Mg-3Y sheet (D, E, F, G, H, I and J are different texture component mark).

3.4. Mechanical Properties

The macro-morphologies of the final as-rolled Mg-3Y alloy sheet are shown in Figure 23a–c. No crack exists in Mg-3Y-R sheet, indicating the ideal rollability. The tensile stress-strain curves of the sheets at room temperature are shown in Figure 23d. The 0.2% yield strength (YS), ultimate tensile strength (UTS), and elongation to failure (FE) are summarized in Table 4. The YS, UTS, and FE of the Mg-R are 142 ± 5 MPa, 196 ± 7 MPa, and 6.4 ± 0.3%, respectively. In contrast, the room temperature YS, UTS, and FE of Mg-RA sheet slightly decrease to 140 ± 3 MPa, 187 ± 5 MPa, and 4.6 ± 0.4%, respectively, indicating worsening strength and ductility. The deterioration of mechanical properties should be attributed to the coarsening of grain during SRX.

Table 4. Tensile properties of the sheets at room temperature.

Sheets	YS (MPa)	UTS (MPa)	FE (%)
Mg-R	142 ± 5	196 ± 7	6.4 ± 0.3
Mg-RA	140 ± 3	187 ± 5	4.6 ± 0.4
Mg-3Y-R	202 ± 4	228 ± 6	18.6 ± 0.6
Mg-3Y-RA	108 ± 6	180 ± 8	25.6 ± 0.8

Compared with Mg-R sheet, the YS and UTS of Mg-3Y-R sheet increase by about 60 and 32 MPa, respectively. The synergistic effect of grain refinement strengthening, second phase strengthening, solid solution strengthening, and dislocation strengthening improves the UTS. The improvement of YS could be contributed to the fine grain strengthening, in accordance with the Hall-Petch relation [61]. As for ductility, the FE of Mg-3Y-R sheet is about three times the size of Mg-R sheet. The weakened double peak texture and the refined grains should be the main reason for the improvement of ductility. As the tensile flow curve shows, the Mg-3Y-RA sheet exhibits a gradual work hardening behavior and a prolonged portion of plastic instability after necking in the tensile curves. Although the YS and UTS of Mg-3Y-RA sheets are decreased to 108 ± 6 MPa and 180 ± 8 MPa, respectively compared with Mg-3Y-R sheet due to the weakening effect of grain refinement strengthening and dislocation strengthening, the FE further increases to 25.6 ± 0.8% after annealing. High ductility of Mg-3Y-RA sheet should be derived from the following four aspects: (1) Significantly reduced kink bands. In tensile process, these existing kink bands would obstruct the subsequent dislocation motion, and easily act as initiation sites for crack nucleation and propagation. This would result in the deterioration of the ductility. (2) Enhanced activity of (0001)<11-20> basal <a> slip. Basal <a> slip could be more easily activated at lower critical resolved shear stress due to the increased average SF values of basal <a> slip. (3) Increased strain hardening behavior. It has been generally accepted that increasing the strain hardening could inhibit the onset of localized deformation, improve the uniform elongation, and therefore increase the elongation to failure [62,63]. (4) Texture weakening. The weakened texture can not only contribute to accommodating the basal slip in tension, but also have a significant impact on the twinning response [62].

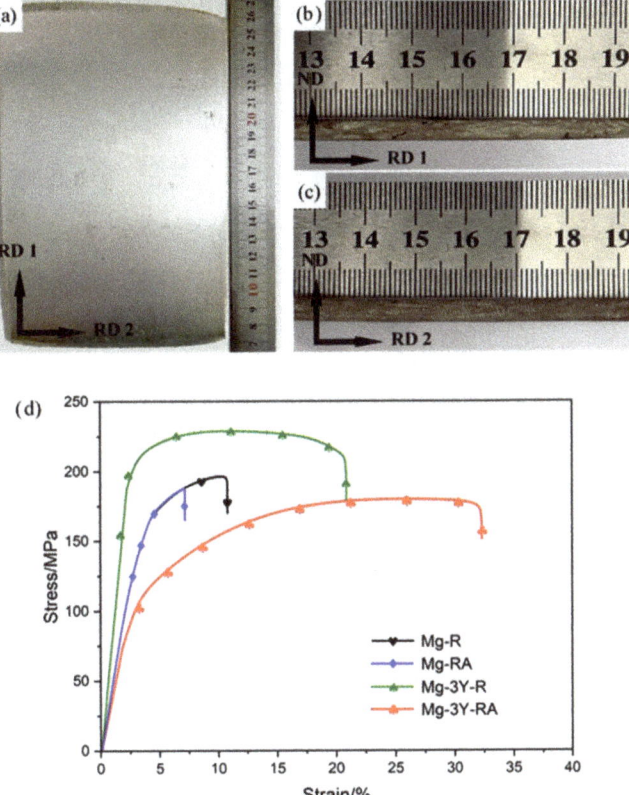

Figure 23. Macro-morphologies of as-rolled Mg-3Y alloy sheet obtained from (**a**) RD1-RD2 plane, (**b**) RD1-ND plane, (**c**) RD2-ND plane, and (**d**) typical engineering stress-strain curves of tensile test for all sheets at room temperature.

The SEM images of tensile fracture surface are presented in Figure 24. Both the Mg-R and Mg-RA sheets exhibit large cleavage facets, which contain tear ridges and a small number of dimples (Figure 24a,b). Therefore, this fracture mode can be categorized as a quasi-cleavage fracture [51]. However, in the Mg-3Y-R and Mg-3Y-RA sheets, a ductile fracture surface with elongated dimples is observed (Figure 24c,e). The cleavage facets are reduced and a tearing characteristic appears in Mg-3Y-RA sheet after annealing. The appearance of tear ridges should be originated from the grain growth during SRX. In addition, the cracked particles inside the dimples (marked by yellow rectangle squares in Figure 24d,f) are observed in both Mg-3Y-R and Mg-3Y-RA sheets. However, the Mg-3Y-RA sheet possesses deeper dimples compared with the as-rolled condition, indicating the highest room temperature ductility among all of the sheets. The change of fracture morphology corresponds well with the improvement of elongation.

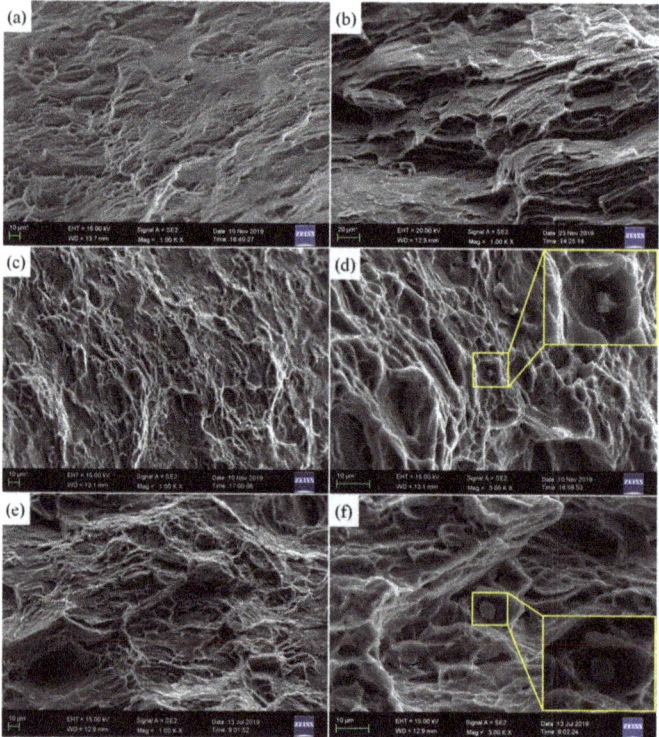

Figure 24. SEM micrographs of the fracture surface of (**a**) as-rolled and (**b**) as-annealed pure Mg sheets; (**c**,**d**) as-rolled and (**e**,**f**) as-annealed Mg-3Y sheets.

3.5. Stretch Formability

The results of Erichsen test are summarized in Table 5. Both Mg-R and Mg-RA sheets show a poor stretch formability, especially Mg-RA sheet. The index Erichsen (IE) value of Mg-3Y-R sheet is 4.18 ± 0.08 mm, which exhibits relatively good formability compared with Mg-R and Mg-RA sheets. The improvement of formability originated from the reduction of basal texture intensity (from 10.4 to 5.2 MRD) and inclination of basal pole (about ±20° tilted to RD2) [15,64–66]. The IE value of Mg-3Y-RA sheet is significantly increased from 4.18 ± 0.08 mm to 6.22 ± 0.05 mm after annealing. The Mg-3Y-RA sheet performs a more significant stretch formability, which is attributed to the further weakened texture. Macromorphology of pure Mg and Mg-3Y sheets after the Erichsen test are shown in Figure 25. Top views of the pure Mg sheets after the Erichsen test reveal that both Mg-R and Mg-RA sheets exhibit a surface crack splitting along multiple directions, which corresponds to the strong butterfly-shaped basal texture (Figure 25c,d). However, the top view of the Mg-3Y-R sheet, which has a basal texture with a splitting of basal pole toward the RD2, exhibits a surface crack parallel to the RD2 after the Erichsen test (Figure 25g). In addition, the circular arc shaped surface crack appears along the angle between RD1 and RD2 in Mg-3Y-RA sheet, which could be related to the randomization of texture (Figure 25h). These results indicate that the splitting of basal pole toward the RD2 and the change of the texture character would strongly affect the macroscopic fracture behaviors of the sheets during Erichsen tests.

Table 5. Stretch formability of the sheets at room temperature.

Sheets	Punch Force (kN)	IE (mm)
Mg-R	1.33 ± 0.03	3.3 ± 0.04
Mg-RA	1.19 ± 0.02	2.8 ± 0.06
Mg-3Y-R	2.02 ± 0.03	4.2 ± 0.08
Mg-3Y-RA	4.61 ± 0.04	6.2 ± 0.05

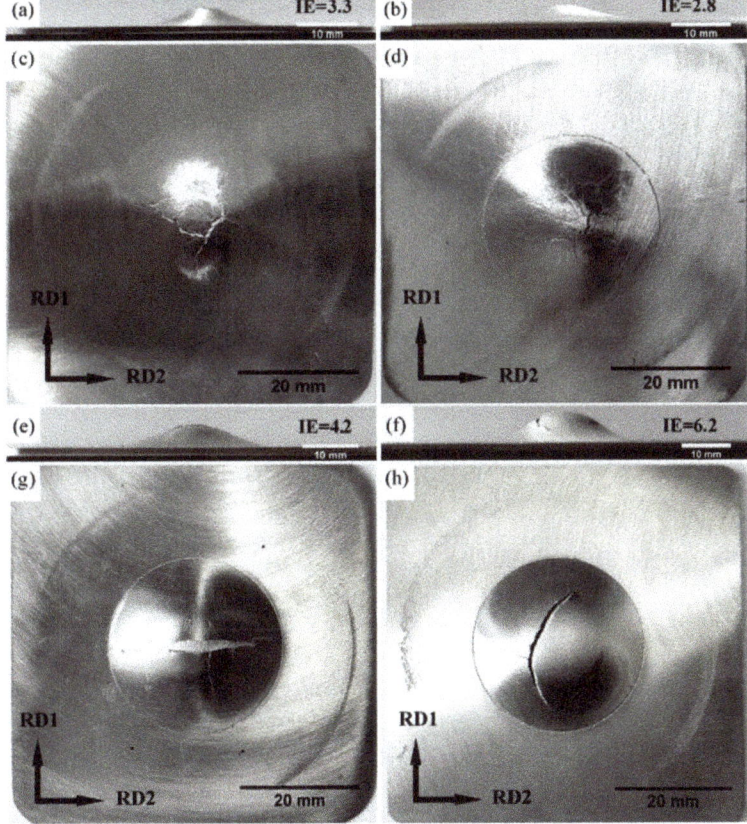

Figure 25. Macro-morphology of specimens after the Erichsen test at room temperature of (**a,c**) as-rolled and (**b,d**) as-annealed pure Mg sheets; (**e,g**) as-rolled and (**f,h**) as-annealed Mg-3Y sheets.

4. Conclusions

In this study, a multi-pass high temperature cross-rolling with inter-pass annealing was applied to Mg-3Y alloy. With pure Mg as a reference, the microstructures, texture, mechanical properties, and stretch formability of Mg-3Y alloy were investigated. The main conclusions are summarized as follows:

(1) The morphological characteristic of β-$Mg_{24}Y_5$ phases with micro-nano size coexistence is formed in Mg matrix, which should be relevant to the solidification segregation behavior of Y solute.

(2) The Mg-3Y-R sheet exhibits a relatively homogeneous deformed microstructure consisting of deformed grains with extensive kink bands and dispersed second phase particles. A double peak texture character appears in Mg-3Y-R sheet with a remarkably reduced pole density and a split of the texture peaks by about $\pm 20°$ tilted to rolling direction 2.

(3) The Mg-3Y-RA sheet presents a complete SRXed microstructure consisting of uniform equiaxed grains. The double texture disappears and a weakened multiple-peak texture appears. The maximum pole density of (0002) basal plane is further decreased from 5.2 to 3.1 MRD. The change of texture that occurs in the Mg-3Y-RA sheet should be due to the strong SRX induced by kink bands and grain boundaries.

(4) Compared with the pure Mg, the Mg-3Y alloy sheet achieved a simultaneous improvement of ductility and stretch formability via high temperature cross-rolling and subsequent short-term annealing. High ductility and stretch formability are attributed to the fine dispersed β-$Mg_{24}Y_5$ phases, homogeneous SRXed microstructure, enhanced activity of basal <a> slip and non-basal slip, and weakening of texture.

Author Contributions: Investigation, writing—review and editing, Y.W.; supervision, review and editing, C.L. and Y.F.; formal analysis, Z.S.; investigation, X.C.; project administration, methodology, review and editing, Y.X. and X.Z. All authors have read and agreed to the published version of the manuscript.

Funding: This work was funded by the Major Special Projects of the Plan "Science and Technology Innovation 2025" in Ningbo (Nos. 2019B10105, 2019B10086, 2020Z096) and Ningbo Natural Science Foundation (No. 202003N4340).

Institutional Review Board Statement: Not applicable.

Informed Consent Statement: Not applicable.

Data Availability Statement: The data presented in this study are available upon request from the corresponding author. The data are not publicly available due to the requirements of related projects.

Acknowledgments: The authors would like to thank all researchers at the Light Metal group of Ningbo Branch of Chinese Academy of Ordnance Science for assistance with data analysis.

Conflicts of Interest: The authors declare they have no known competing financial interests or personal relationships that could have influenced the work reported in this paper.

References

1. Kainer, K.U. *Magnesium Alloys and Technologies*; Wiley-VCH Verlag GmbH & Co.: Weinheim, Germany, 2003.
2. Ion, S.E.; Humphreys, F.J.; White, S.H. Dynamic recrystallization and the development of microstructure during the high temperature deformation of magnesium. *Acta Metall.* **1982**, *30*, 1909–1919. [CrossRef]
3. Humphreys, F.J.; Hatherly, M. *Recrystallization and Related Annealing Phenomena*; Pergamon Press: Oxford, UK, 1996.
4. Pekguleryuz, M.O.; Kainer, K.U.; Arslan Kaya, A. *Fundamentals of Magnesium Alloy Metallurgy*; Woodhead Publishing: Cambridge, UK, 2013.
5. Yang, X.; Miura, H.; Sakai, T. Dynamic evolution of new grains in magnesium alloy AZ31 during hot deformation. *Mater. Trans.* **2003**, *44*, 197–203. [CrossRef]
6. Yang, X.; Ji, Z.; Miura, H.; Sakai, T. Dynamic recrystallization and texture development during hot deformation of magnesium alloy AZ31. *Trans. Nonferrous Met. Soc. China* **2009**, *19*, 55–60. [CrossRef]

7. Agnew, S.R.; Yoo, M.H.; Tomé, C.N. Application of texture simulation to understanding mechanical behavior of Mg and solid solution alloys containing Li or Y. *Acta Mater.* **2001**, *49*, 4277–4289. [CrossRef]
8. Barnett, M.R.; Nave, M.D.; Bettles, C.J. Deformation microstructures and textures of some cold rolled Mg alloys. *Mater. Sci. Eng. A* **2004**, *386*, 205–211. [CrossRef]
9. Jager, A.; Lukac, P.; Gartnerova, V.; Haloda, J.; Dopita, M. Influence of annealing on the microstructure of commercial Mg alloy AZ31 after mechanical forming. *Mater. Sci. Eng. A* **2006**, *432*, 20–25. [CrossRef]
10. Yang, X.Y.; Okabe, Y.; Miura, H.; Sakai, T. Annealing of a magnesium alloy AZ31 after interrupted cold deformation. *Mater. Des.* **2012**, *36*, 626–632. [CrossRef]
11. Bohlen, J.; Nürnberg, M.R.; Senn, J.W.; Letzig, D.; Agnew, S.R. The texture and anisotropy of magnesium-zinc-rare earth alloy sheets. *Acta Mater.* **2007**, *55*, 2101–2112. [CrossRef]
12. Hantzsche, K.; Bohlen, J.; Wendt, J.; Kainer, K.U.; Yi, S.B.; Letzig, D. Effect of rare earth additions on microstructure and texture development of magnesium alloy sheets. *Scr. Mater.* **2010**, *63*, 725–730. [CrossRef]
13. Stanford, N. Micro-alloying Mg with Y, Ce, Gd and La for texture modification-A comparative study. *Mater. Sci. Eng. A* **2010**, *527*, 2669–2677. [CrossRef]
14. Stanford, N.; Barnett, M. Effect of composition on the texture and deformation behaviour of wrought Mg alloys. *Scr. Mater.* **2008**, *58*, 179–182. [CrossRef]
15. Huang, X.S.; Suzuki, K.; Watazu, A.; Shigematsu, I.; Saito, N. Improvement of formability of Mg-Al-Zn alloy sheet at low temperatures using differential speed rolling. *J. Alloys Compd.* **2009**, *470*, 263–268. [CrossRef]
16. Chino, Y.; Mabuchi, M.; Kishihara, R.; Hosokawa, H.; Yamada, Y.; Wen, C.E.; Shimojima, K.; Iwasaki, H. Mechanical properties and press formability at room temperature of AZ31 Mg alloy processed by single roller drive rolling. *Mater. Trans.* **2002**, *43*, 2554–2560. [CrossRef]
17. Chino, Y.; Sassa, K.; Kamiya, A.; Mabuchi, M. Microstructure and press formability of a cross-rolled magnesium alloy sheet. *Mater. Lett.* **2007**, *61*, 1504–1506. [CrossRef]
18. Song, B.; Huang, G.S.; Li, H.C.; Zhang, L.; Huang, G.J.; Pan, F.S. Texture evolution and mechanical properties of AZ31B magnesium alloy sheets processed by repeated unidirectional bending. *J. Alloys Compd.* **2010**, *489*, 475–481. [CrossRef]
19. Sunaga, Y.; Tanaka, Y.; Asakawa, M.; Katoh, M.; Kobayashi, M. Effect of twin formation by repetitive bending on texture of AZ61 magnesium alloy sheet and improvement of formability. *J. Jpn. Inst. Light Met.* **2009**, *59*, 655–658. [CrossRef]
20. Cheng, Y.Q.; Chen, Z.H.; Xia, W.J. Drawability of AZ31 magnesium alloy sheet produced by equal channel angular rolling at room temperature. *Mater. Charact.* **2007**, *58*, 617–622. [CrossRef]
21. Yamamoto, A.; Tsukahara, Y.; Fukumoto, S. Effects of wavy roll-forming on textures in AZ31B magnesium alloy. *Mater. Trans.* **2008**, *49*, 995–999. [CrossRef]
22. Kohzu, M.; Kii, K.; Nagata, Y.; Nishio, H.; Higashi, K.; Inoue, H. Texture randomization of AZ31 magnesium alloy sheets for improving the cold formability by a combination of rolling and high-temperature annealing. *Mater. Trans.* **2010**, *51*, 749–755. [CrossRef]
23. Huang, X.S.; Suzuki, K.; Saito, N. Enhancement of stretch formability of Mg-3Al-1Zn alloy sheet using hot rolling at high temperatures up to 823 K and subsequent warm rolling. *Scr. Mater.* **2009**, *61*, 445–448. [CrossRef]
24. Huang, X.S.; Suzuki, K.; Chino, Y. Influences of initial texture on microstructure and stretch formability of Mg-3Al-1Zn alloy sheet obtained by a combination of high temperature and subsequent warm rolling. *Scr. Mater.* **2010**, *63*, 395–398. [CrossRef]
25. Chino, Y.; Mabuchi, M. Enhanced stretch formability of Mg-Al-Zn alloy sheets rolled at high temperature (723 K). *Scr. Mater.* **2009**, *60*, 447–450. [CrossRef]
26. Huang, X.S.; Suzuki, K.; Saito, N. Textures and stretch formability of Mg-6Al-1Zn magnesium alloy sheets rolled at high temperatures up to 793 K. *Scr. Mater.* **2009**, *60*, 651–654. [CrossRef]
27. Huang, X.S.; Suzuki, K.; Chino, Y.; Mabuchi, M. Improvement of stretch formability of Mg-3Al-1Zn alloy sheet by high temperature rolling at finishing pass. *J. Alloys Compd.* **2011**, *509*, 7579–7584. [CrossRef]
28. Huang, X.; Suzuki, K.; Chino, Y.; Mabuchi, M. Texture and stretch formability of AZ61 and AM60 magnesium alloy sheets processed by high-temperature rolling. *J. Alloys Compd.* **2015**, *632*, 94–102. [CrossRef]
29. Huang, X.S.; Suzuki, K.; Chino, Y.; Mabuchi, M. Influence of initial texture on cold deep drawability of Mg-3Al-1Zn alloy sheets. *Mater. Sci. Eng. A* **2013**, *565*, 359–372. [CrossRef]
30. Beausir, B.; Biswas, S.; Kim, D.I.; Tóth, L.S.; Suwas, S. Analysis of microstructure and texture evolution in pure magnesium during symmetric and asymmetric rolling. *Acta Mater.* **2009**, *57*, 5061–5077. [CrossRef]
31. Chino, Y.; Sassa, K.; Kamiya, A.; Mabuchi, M. Stretch formability at elevated temperature of a cross-rolled AZ31 Mg alloy sheet with different rolling routes. *Mater. Sci. Eng. A* **2008**, *473*, 195–200. [CrossRef]
32. Chen, T.; Chen, Z.Y.; Yi, L.; Xiong, J.Y.; Liu, C.M. Effects of texture on anisotropy of mechanical properties in annealed Mg-0.6%Zr–1.0%Cd sheets by unidirectional and cross rolling. *Mater. Sci. Eng. A* **2014**, *615*, 324–330. [CrossRef]
33. Brown, D.W.; Agnew, S.R.; Bourke, M.A.M.; Holden, T.M.; Vogel, S.C.; Tomé, C.N. Internal strain and texture evolution during deformation twinning in magnesium. *Mater. Sci. Eng. A* **2005**, *399*, 1–12. [CrossRef]

34. Li, X.; Al-Samman, T.; Gottstein, G. Mechanical properties and anisotropy of ME20 magnesium sheet produced by unidirectional and cross rolling. *Mater. Des.* **2011**, *32*, 4385–4393. [CrossRef]
35. Zhang, J.; Mao, C.; Long, C.G.; Chen, J.; Tang, K.; Zhang, M.J.; Peng, P. Phase stability, elastic properties and electronic structures of Mg-Y intermetallics from first-principles calculations. *J. Magnes. Alloy.* **2015**, *3*, 127–133. [CrossRef]
36. StJohn, D.H.; Qian, M.; Easton, M.A.; Cao, P.; Hildebrand, Z. Grain refinement of magnesium alloys. *Metall. Mater. Tran. A* **2005**, *36*, 1669–1679. [CrossRef]
37. Lee, Y.C.; Dahle, A.K.; StJohn, D.H. The role of solute in grain refinement of magnesium. *Metall. Mater. Tran. A* **2000**, *31*, 2895–2906. [CrossRef]
38. Shi, B.Q.; Chen, R.S.; Ke, W. Effects of yttrium and zinc on the texture, microstructure and tensile properties of hot-rolled magnesium plates. *Mater. Sci. Eng. A* **2013**, *560*, 62–70. [CrossRef]
39. Yu, Z.J.; Xu, C.; Meng, J.; Zhang, X.H.; Kamadoa, S. Microstructure evolution and mechanical properties of as-extruded Mg-Gd-Y-Zr alloy with Zn and Nd additions. *Mater. Sci. Eng. A.* **2018**, *713*, 234–243. [CrossRef]
40. Zhang, L.; Wang, Q.D.; Liu, G.P.; Guo, W.; Jiang, H.Y.; Ding, W.J. Effect of SiC particles and the particulate size on the hot deformation and processing map of AZ91 magnesium matrix composites. *Mater. Sci. Eng. A.* **2017**, *707*, 315–324. [CrossRef]
41. Christian, J.W.; Mahajan, S. Deformation twinning. *Prog. Mater. Sci.* **1995**, *39*, 1–157. [CrossRef]
42. Yang, X.Y.; Jiang, Y.P. Morphology and crystallographic characteristics of deformation bands in Mg alloy under hot deformation. *Acta Metall. Sin.* **2010**, *46*, 451–457. [CrossRef]
43. Zhou, B.; Sui, M.L. Generation and interaction mechanism of tension kink band in AZ31 magnesium alloy. *Acta Metall. Sin.* **2019**, *55*, 1512–1518.
44. Matsumoto, T.; Yamasaki, M.; Hagihara, K.; Kawamura, Y. Configuration of dislocations in low-angle kink boundaries formed in a single crystalline long-period stacking ordered Mg-Zn-Y alloy. *Acta Mater.* **2018**, *151*, 112–124. [CrossRef]
45. Hong, S.G.; Park, S.H.; Lee, C.S. Role of {10-12} twinning characteristics in the deformation behavior of a polycrystalline magnesium alloy. *Acta Mater.* **2010**, *58*, 5873–5885. [CrossRef]
46. Al-Samman, T.; Gottstein, G. Room temperature formability of a magnesium AZ31 alloy: Examining the role of texture on the deformation mechanisms. *Mater. Sci. Eng. A* **2008**, *488*, 406–414. [CrossRef]
47. Wright, S.I.; Nowell, M.M.; Field, D.P. A review of strain analysis using electron backscatter diffraction. *Microsc. Microanal.* **2011**, *17*, 316–329. [CrossRef]
48. Kim, Y.J.; Kim, S.H.; Lee, J.U.; You, B.S.; Park, S.H. Evolution of high-cycle fatigue behavior of extruded AZ91 alloy by artificial cooling during extrusion. *Mater. Sci. Eng. A* **2017**, *707*, 620–628. [CrossRef]
49. Al-Samman, T.; Molodov, K.D.; Molodov, D.A.; Gottstein, G.; Suwas, S. Softening and dynamic recrystallization in magnesium single crystals during c-axis compression. *Acta Mater.* **2012**, *60*, 537–545. [CrossRef]
50. Yang, Y.; Yang, X.; Xiao, Z.; Zhang, D.; Wang, J.; Sakai, T. Annealing behavior of a cast Mg-Gd-Y-Zr alloy with necklace fine grains developed under hot deformation. *Mater. Sci. Eng. A* **2017**, *688*, 280–288. [CrossRef]
51. Wu, W.X.; Jin, L.; Wang, F.H.; Sun, J.; Zhang, Z.Y.; Ding, W.J.; Dong, J. Microstructure and texture evolution during hot rolling and subsequent annealing of Mg-1Gd alloy. *Mater. Sci. Eng. A* **2013**, *582*, 194–202. [CrossRef]
52. Sandlöbes, S.; Friák, M.; Zaefferer, S.; Dick, A.; Yi, S.; Letzig, D.; Pei, Z.; Zhu, L.F.; Neugebauer, J.; Raabe, D. The relation between ductility and stacking fault energies in Mg and Mg-Y alloys. *Acta Mater.* **2012**, *60*, 3011–3021. [CrossRef]
53. Sandlöbes, S.; Zaefferer, S.; Schestakow, I.; Yi, S.; Gonzalez-Martinez, R. On the role of non-basal deformation mechanisms for the ductility of Mg and Mg-Y alloys. *Acta Mater.* **2011**, *59*, 429–439. [CrossRef]
54. Hirsch, J.; Al-Samman, T. Superior light metals by texture engineering: Optimized aluminum and magnesium alloys for automotive applications. *Acta Mater.* **2013**, *61*, 818–843. [CrossRef]
55. Miao, Q.; Hu, L.X.; Wang, G.J.; Wang, E.R. Fabrication of excellent mechanical properties AZ31 magnesium alloy sheets by conventional rolling and subsequent annealing. *Mater. Sci. Eng. A* **2011**, *528*, 6694–6701. [CrossRef]
56. Agnew, S.R.; Duygulu, Ö. Plastic anisotropy and the role of non-basal slip in magnesium alloy AZ31B. *Int. J. Plast.* **2005**, *21*, 1161–1193. [CrossRef]
57. Wusatowska-Sarnek, A.M.; Miura, H.; Sakai, T. Influence of deformation temperature on microstructure evolution and static recrystallization of polycrystalline copper. *Mater. Trans.* **2001**, *42*, 2452–2459. [CrossRef]
58. Huang, X.S.; Suzuki, K.; Chino, Y. Annealing behaviour of Mg-3Al-1Zn alloy sheet obtained by a combination of high-temperature rolling and subsequent warm rolling. *J. Alloys Compd.* **2011**, *509*, 4854–4860. [CrossRef]
59. Suh, B.C.; Shim, M.S.; Shin, K.S.; Kim, N.J. Current issues in magnesium sheet alloys: Where do we go from here? *Scr. Mater.* **2014**, *84–85*, 1–6. [CrossRef]
60. Bian, M.Z.; Sasaki, T.T.; Suh, B.C.; Nakata, T.; Kamado, S.; Hono, K. A heat-treatable Mg-Al-Ca-Mn-Zn sheet alloy with good room temperature formability. *Scr. Mater.* **2017**, *138*, 151–155. [CrossRef]
61. Zhang, B.; Peng, X.D.; Ma, Y.; Li, Y.M.; Yu, Y.Q.; Wei, G.B. Microstructure and mechanical properties of Mg-9Li-3Al-xGd alloys. *Mater. Sci. Technol.* **2015**, *31*, 1035–1041. [CrossRef]
62. Stanford, N.; Barnett, M.R. The origin of "rare earth" texture development in extruded Mg-based alloys and its effect on tensile ductility. *Mater. Sci. Eng. A* **2008**, *496*, 399–408. [CrossRef]

63. Stanford, N.; Atwell, D.; Barnett, M.R. The effect of Gd on the recrystallisation, texture and deformation behaviour of magnesium-based alloys. *Acta Mater.* **2010**, *58*, 6773–6783. [CrossRef]
64. Yukutake, E.; Kaneko, J.; Sugamata, M. Anisotropy and non-uniformity in plastic behavior of AZ31 magnesium alloy plates. *Mater. Trans.* **2003**, *44*, 452–457. [CrossRef]
65. Iwanaga, K.; Tashiro, H.; Okamoto, H.; Shimizu, K. Improvement of formability from room temperature to warm temperature in AZ31 magnesium alloy. *J. Mater. Process. Technol.* **2004**, *155–156*, 1313–1316. [CrossRef]
66. Chino, Y.; Sassa, K.; Mabuchi, M. Texture and stretch formability of Mg-1.5 mass% Zn-0.2 mass% Ce alloy rolled at different rolling temperatures. *Mater. Trans.* **2008**, *49*, 2916–2918. [CrossRef]

Article

Study on Microstructural Evolution and Mechanical Properties of Mg-3Sn-1Mn-xLa Alloy by Backward Extrusion

Xuefei Zhang [1], Baoyi Du [1] and Yuejie Cao [2,*]

[1] School of Mechanical Engineering, Shenyang University, Shenyang 110044, China
[2] School of Aeronautics, Chongqing Jiaotong University, Chongqing 400074, China
* Correspondence: caoyj@cqjtu.edu.cn

Abstract: Mg-3Sn-1Mn-xLa alloy bars were prepared using backward extrusion, and the effects of the La content on the microstructures and mechanical properties of the alloy were systematically studied using an optical microscope (OM), X-ray diffraction (XRD), scanning electron microscope (SEM), transmission electron microscope (TEM), and tensile tests. The results of this research show that the Mg_2Sn phases were mainly formed at the α-Mg grain boundaries and within the grains in the Mg-3Sn-1Mn alloy. After adding a certain amount of La, the plate-shaped MgSnLa compounds consisting of $Mg_{17}La_2$, Mg_2Sn, and La_5Sn_3 gradually disappeared in the α-Mg matrix and grain boundaries. With an increase in La content, the Mg_2Sn phase in the crystal was gradually refined and spheroidized. When the content of La reached 1.5%, the tensile strength of the alloy reached 300 Mpa and the elongation reached 12.6%, i.e., 25% and 85% increases, respectively, compared to the Mg-3Sn-1Mn alloy. The plate-shaped compound of Mg-3Sn-1Mn-1.5La had an average length of 3000 ± 50 nm, while the width was 350 ± 10 nm. Meanwhile, the extruded alloy's grain size was significantly refined, and there were many small cleavage steps and dimples in the fracture surface of the alloy. When the La content reached 2%, the alloy performance showed a downward trend due to the coarsening of the grains. The formed plate-shaped MgSnLa compounds and Mg_2Sn phases were consistent with the α-Mg matrix. They effectively pinned the dislocations and grain boundaries, which is the main reason for strengthening the mechanical properties of extrusion alloys.

Keywords: Mg alloy; backward extrusion; tensile strength; failure mechanism

Citation: Zhang, X.; Du, B.; Cao, Y. Study on Microstructural Evolution and Mechanical Properties of Mg-3Sn-1Mn-xLa Alloy by Backward Extrusion. *Materials* **2023**, *16*, 4588. https://doi.org/10.3390/ma16134588

Academic Editor: Hajo Dieringa

Received: 9 May 2023
Revised: 9 June 2023
Accepted: 12 June 2023
Published: 25 June 2023

Copyright: © 2023 by the authors. Licensee MDPI, Basel, Switzerland. This article is an open access article distributed under the terms and conditions of the Creative Commons Attribution (CC BY) license (https://creativecommons.org/licenses/by/4.0/).

1. Introduction

Magnesium and its alloys have the advantages of low density, high specific strength, and stiffness and are widely used in aerospace, national defense, the military industry, the automotive industry, and other fields. However, the low plasticity at room temperature, poor heat, and poor corrosion resistance of traditional magnesium alloys limits their wide application [1–5]. Alloying is an effective method to improve the mechanical properties of magnesium alloys [6]. Previous studies have shown that, when Sn is added to magnesium, the Mg_2Sn phase formed has a high melting point and high hardness, which can effectively improve the mechanical properties and thermal stability of the alloy, as well as improve the high-temperature strength and creep resistance of magnesium alloys [7–12]. In addition, rare earth strengthening is also one of the methods that can effectively improve the mechanical properties of magnesium alloys, because adding rare earth can improve the deformation ability and strength of magnesium alloys through the precipitation hardening and solid solution strengthening mechanisms [13–15]. Current research results indicate that adding rare earth elements such as La, Ce, Ga, Y, and Nd to magnesium alloys can effectively refine the α-Mg grains and improve the microstructure of the alloy, thereby significantly improving the comprehensive mechanical properties of the magnesium alloys [16–19]. Zengin et al. [18] studied the mechanism of the effect of La on the grain refinement process of magnesium alloys. They found that, during the solidification process of the alloy, the enrichment of La atoms at the solid–liquid interface accelerated the nucleation rate while inhibiting the growth of dendrites.

Grain refinement is a critical way to improve the plasticity and toughness of magnesium alloys [5]. In addition to adding rare earths, high-temperature thermoplastic deformation is also a very effective method. High-temperature large plastic deformation can not only effectively eliminate internal defects and voids in as-cast magnesium alloys but also plays a significant role in grain refinement, improving the alloy microstructure's uniformity and consistency [20]. The grain refinement after extrusion is attributed to typical dynamic recrystallization (DRX) occurring during thermoplastic deformation [21,22]. During plastic deformation, high-density dislocations are introduced into coarse grains, which aggregate to promote the formation of low-angle grain boundaries (LAGBs). As the strain increases, a transition occurs from small-angle grain boundaries to high-angle grain boundaries (HAGBs), resulting in dynamic recrystallization (DRX). During dynamic recrystallization, precipitates can pin dislocations and grain boundaries, inhibiting grain growth [11,23]. Some studies have reported that, on the one hand, precipitated phases bind dislocations, increasing the driving force for recrystallization and promoting dynamic recrystallization (DRX). On the other hand, precipitates bind low-angle grain boundaries (LAGBs) and grain boundaries generated by dynamic crystallization (DRX), inhibiting the transition from low-angle grain boundaries to high-angle grain boundaries (HAGBs), thereby delaying dynamic recrystallization [24–27]. Furthermore, other studies have also found that a series of alloys that have undergone alloying and rare earth alloying, such as Mg-Sn-Ca-Mn [13,28,29], Mg-Sn-Mn [30,31], Mg-Sn-Zn-Mn [32–34], and Mg-Gd-Y-Zn [35,36], exhibit excellent mechanical properties at high temperatures and at room temperature.

Magnesium alloys with a hexagonal close-packed (HCP) crystal structure are difficult to process at room temperature. Currently, they are mainly prepared by casting and thermoforming processes. Based on the above research, Zhao et al. [11,37,38] designed a Mg-3Sn-1Mn-1La alloy using a continuous rheo-rolling method. The Mg_2Sn phase mainly exists in the Mg-3Sn-1Mn alloy without adding La. After adding a certain amount of La, MgSnLa metal compounds composed of the La_5Sn_3, Mg_2Sn, and $Mg_{17}La_2$ phases are generated. The presence of these metal compounds improves the room-temperature and high-temperature properties of magnesium alloys. Magnesium alloys can also achieve grain refinement through hot-extrusion deformation and precipitation, which positively impact the alloy's microstructure and comprehensive mechanical properties [18,26]. However, reports have yet to be made on preparing rare earth magnesium alloys by backward extrusion and their strengthening effects on the microstructure and properties.

In this study, the effects of La on the microstructure and properties of hot-formed Mg-3Sn-1Mn-xLa alloy wire produced by the backward-extrusion process were studied. The objective of the study described here is to reveal the effect of the precipitates on grain refinement during the hot-forming process and to elucidate the effect of the rare earth phase on the mechanical properties of the Mg-3Sn-1Mn-xLa alloy. The results and related discussions will provide an important basis for understanding the fine-grain strengthening mechanism of Mg-3Sn-1Mn alloys and developing high-performance La-containing magnesium alloys.

2. Experimental Procedure

Mg-3Sn-1Mn-xLa alloys were prepared by melting pure Mg (99.99% purity), pure tin (99.9% purity), Mg-75Mn (wt.%), and Mg-30La (wt.%). Pure Mg and all other raw materials were preheated in the oven to eliminate water vapor. A resistance furnace (3 kW, SG2-3-9, Shenyang General Furnace Manufacturing Co., Ltd., Shenyang, China) was used to melt Mg alloys. The magnesium ingots were then melted in a clay graphite crucible using an electric resistance furnace at 700 °C to ensure complete homogenization, and the dried Sn, Mn, and Mg-La master alloys were added in sequence. After mechanical stirring, the gas and slag were removed. The melts were then degassed with argon and poured into a low-carbon steel mold of 180 mm height and 40 mm diameter. The chemical composition of the Mg alloy is shown in Table 1.

Table 1. Detailed composition of Mg-Sn-Mn-xLa alloys.

Number	Nominal Alloy	Sn (wt.%)	Mn (wt.%)	La (wt.%)	Mg
#1	Mg-3Sn-1Mn	2.81	0.90	0	Bal
#2	Mg-3Sn-1Mn+0.5La	2.79	0.89	0.46	Bal
#3	Mg-3Sn-1Mn+1.0La	2.83	0.91	0.94	Bal
#4	Mg-3Sn-1Mn+1.5La	2.88	0.93	1.45	Bal
#5	Mg-3Sn-1Mn+2.0La	2.85	0.90	1.92	Bal

The backward-extrusion experiment in this paper used a 300 T single-column vertical hydraulic compressor to prepare extruded samples. The schematic diagram of its working principle is shown in Figure 1. Before extrusion processing, the magnesium alloy as-cast samples were homogenized at 390 °C for 6 h [29,34], while the extrusion cylinder was preheated to 350 °C. Afterward, backward extrusion was carried out at 380 °C to adjust the bar diameter from 40 mm to 10 mm, with an extrusion ratio of 16.

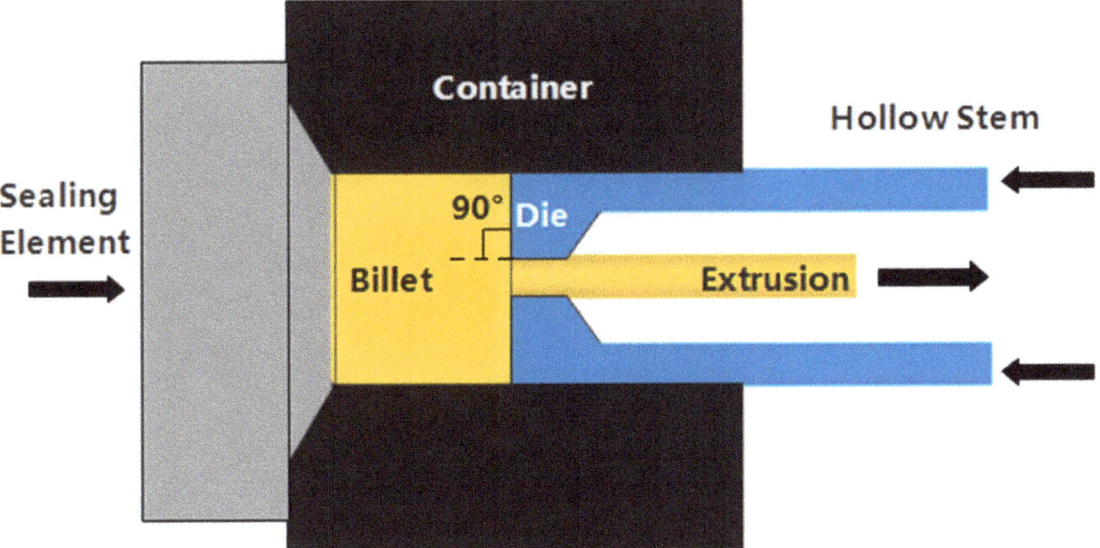

Figure 1. Schematic diagram of the backward-extrusion process.

The mechanical properties of the extruded specimens were tested on a SANS electronic universal tensile testing machine. The geometric dimension of the tensile specimens was designed according to the GB/T 16865-2013 standard [39], and the detailed sample size is shown in Figure 2. The phase identification of the surface was performed using X-ray diffraction (XRD). The specimens were polished and etched with 15 mL HCl + 56 mL C_2H_5OH + 47 mL H_2O. The microstructure observation and microanalysis were performed using an OLYMPUS DSX500 optical microscope and a Hitachi S-4800 II scanning electron microscope. The TEM observations were performed using a field-emission-gun (FEG) Tecnai G^2 20 microscope operating at an accelerating voltage of 200 kV.

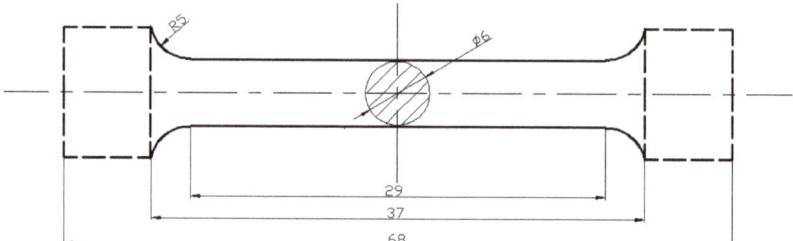

Figure 2. Configuration of the samples used for the tensile tests (unit: mm).

3. Results and Discussions

3.1. Microstructure of As-Cast Alloys

Figure 3 shows the as-cast metallographic microstructure of Mg-3Sn-1Mn-xLa alloys, mainly composed of primary α-Mg and secondary phases. The α-Mg in the alloy specifically presented a dendritic structure, and a large number of grayish-white near-dendritic forms were α-Mg matrix. At the same time, the secondary phase was black and mainly distributed along the grain boundaries. The as-cast structure of the alloy without the addition of La was primarily composed of coarse α-Mg dendrites, interdendrites, and intermetallic compounds between the dendritic arms. The average grain size of the alloy was 150 ± 5 μm; its distribution is shown in Figure 3a. After the addition of La, a large number of black intermetallic compounds appeared along the grain boundaries of the alloy (Figure 3b). With the increase in La content, the number of eutectic compounds increased gradually, and the dendrites started to show refinement (Figure 3c), which was mainly because the second-phase particles formed during solidification prevented the further growth of the dendrites, thus developing finer dendrites. When the La content reached 1.5%, the refinement of the alloy's dendrites was most obvious, and the second phase near the grain boundary increased significantly. The average size of the grains showed a gradual increase to 50 ± 2 μm. The distribution was relatively uniform, which is due to the enrichment of La atoms at the solid–liquid interface during solidification leading to the acceleration of the nucleation rate and the restriction of dendrite growth (Figure 3d). When the rare earth content continued to increase, the excessive enrichment of the precipitated phases increased the segregation, and the grains appeared coarsened instead. The average grain size of the alloy increased to 100 ± 3 μm, as shown in Figure 3e. A more detailed display and analysis of the precipitates will be further introduced in the following sections.

Figure 4 shows the effect of La concentration on precipitate formation in the Mg-3Sn-1Mn-xLa (wt.%) alloy as studied by X-ray diffraction for different La concentrations. It indicates that the sample was mainly composed of α-Mg matrix, and the second-phase Mg_2Sn was only detected in the Mg-3Sn-1Mn alloy (curve (1)). However, with the addition of La (curve (2)), the diffraction peaks of the Mg_2Sn phase began to weaken, indicating that with the increase in La concentration, Sn preferentially formed new phases with La, resulting in a decrease in the Mg_2Sn phases. As shown in curve (4) in Figure 4, when the La concentration increased to 1.5 wt.%, the diffraction peak intensities associated with the Mg_2Sn phase were further decreased. Furthermore, when the La concentration was 0.5 wt.%, the diffraction peaks of the La_5Sn_3 phase and $Mg_{17}La_2$ phase appeared. As the La concentration increased to 1.5 wt.%, the diffraction peak of the La_5Sn_3 phase increased and began to appear at a new peak position. At the same time, the diffraction peaks of the $Mg_{17}La_2$ phase also appeared at a new position, indicating a further increase in the $Mg_{17}La_2$ phase. Generally speaking, the difference in the diffraction peak intensity of Mg-3Sn-1Mn-xLa manifests that the amount of La_5Sn_3 and $Mg_{17}La_2$ precipitates increased apparently with increasing La content.

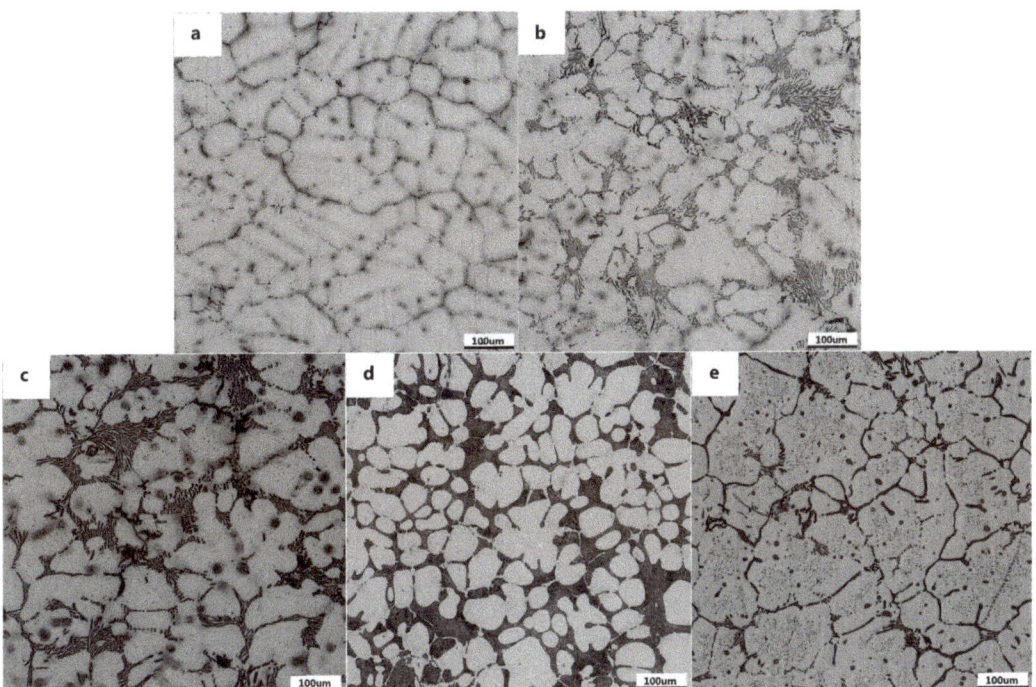

Figure 3. Optical images of the microstructures of the as-cast magnesium alloy: (**a**) Mg-3Sn-1Mn, (**b**) Mg-3Sn-1Mn+0.5La, (**c**) Mg-3Sn-1Mn+1.0La, (**d**) Mg-3Sn-1Mn+1.5La, (**e**) Mg-3Sn-1Mn+2.0La.

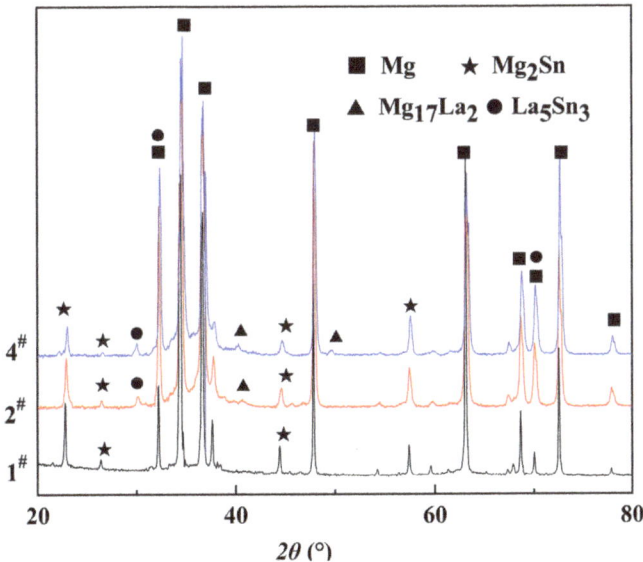

Figure 4. XRD patterns for the Mg-3Sn-1Mn-xLa alloys with 1#, 2#, and 4#.

Figure 5 shows the as-cast SEM microstructure and EDS spectrum analysis of Mg-3Sn-1Mn and Mg-3Sn-1Mn-1.5La alloys. It can be seen from Figure 5a,c that there was an irregularly distributed second phase near the grain boundary in the Mg-3Sn-1Mn alloy. Combined with the XRD phase analysis in Figure 4, the second phase was mainly the Mg$_2$Sn phase. In the Mg-3Sn-1Mn-1.5La alloy, a large number of plate-shaped compounds were distributed at the grain boundaries, and a small number of small, elongated phases were also present in the grains. Through the XRD phase analysis results in Figure 4, it can be seen that the plate-shaped compounds comprised a ternary mixture of La$_5$Sn$_3$, Mg$_2$Sn, and Mg$_{17}$La$_2$ phases.

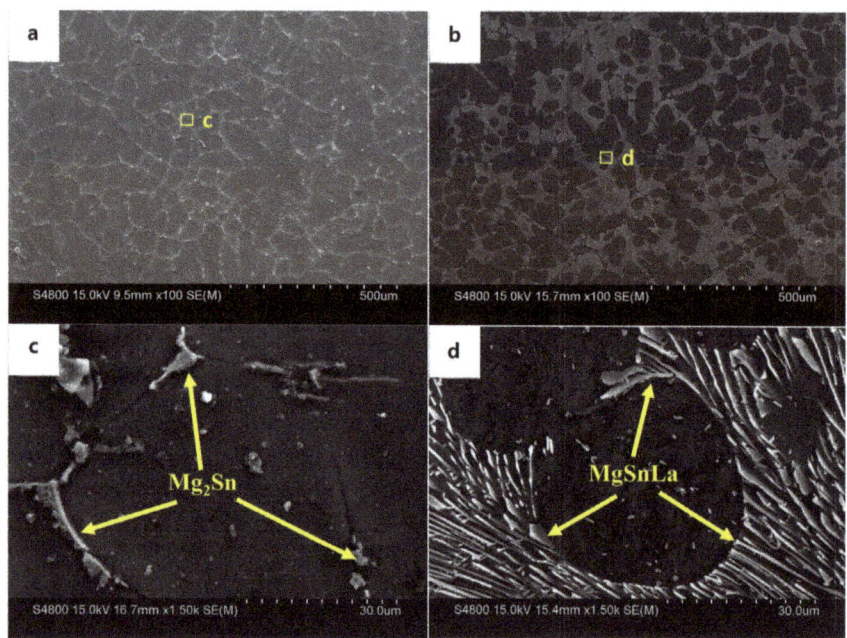

Figure 5. SEM microstructures of as-cast alloys: (**a,c**) Mg-3Sn-1Mn, (**b,d**) Mg-3Sn-1Mn+1.5La.

3.2. Microstructure Morphology of Backward-Extrusion Alloy

The microstructure morphology of Mg-3Sn-1Mn and Mg-3Sn-1Mn-1.5La alloys after backward extrusion in the transverse direction (TD) are shown in Figure 6a,b. Compared with the microstructure of the as-cast Mg alloy in Figure 4, the microstructure had undergone significant changes, with substantial grain refinement and many precipitates appearing on the Mg matrix. The main reason is that Mg alloys are prone to dynamic recrystallization during the hot-forming process to form new recrystallized grains. The accumulation of dislocations at the grain boundaries or second-phase particles of dynamic recrystallization grains can promote the occurrence of dynamic recrystallization [23,40,41]. Furthermore, the microstructure morphology of Mg-3Sn-1Mn and Mg-3Sn-1Mn-1.5La alloys in the extrusion direction (ED) is shown in Figure 6c,d. As can be seen from Figure 6c, the grain structure of the alloy was significantly refined after backward extrusion and typical refined equiaxed grains appeared. This is mainly attributed to the dynamic recrystallization and heterogeneous nucleation of recrystallized grains of the Mg$_2$Sn phase precipitated during the extrusion process and the hindering effect on grain growth. Previous studies have shown that equiaxed grains with a uniform distribution can hinder the formation of the void, thus increasing the elongation to tensile failure [1]. Compared to Mg-3Sn-1Mn alloy, the Mg-3Sn-1Mn-1.5La alloy exhibited many black fibrous structures with streamline distribution along the extrusion direction, which probably resulted from the extrusion deformation of precipitates.

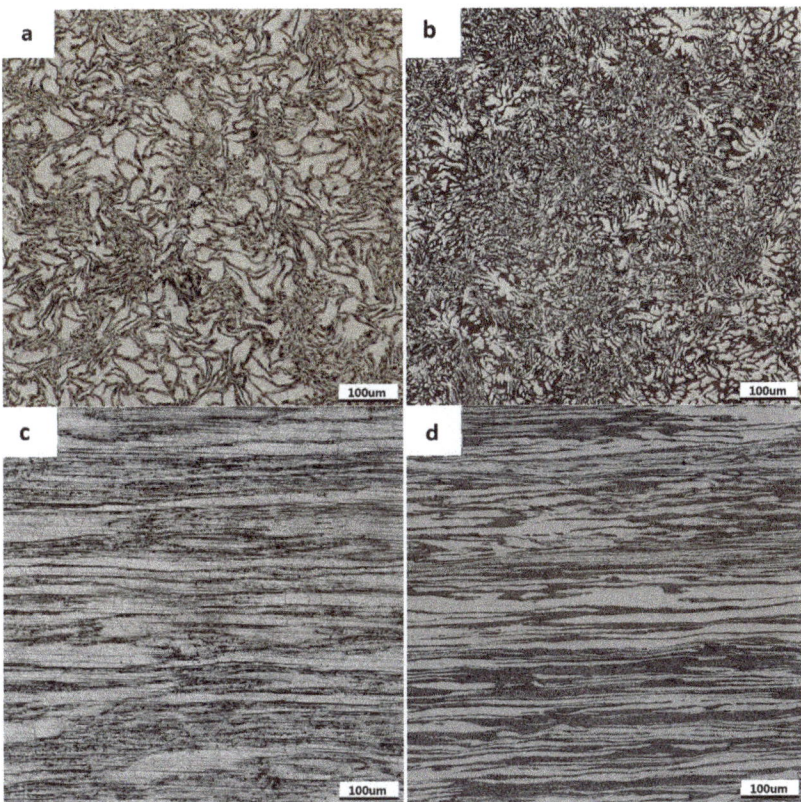

Figure 6. Optical microstructures of Mg-Sn-Mn-xLa alloys backward extruded in transverse and extruded. (**a**,**c**) Mg-3Sn-1Mn (**b**,**d**) Mg-3Sn-1Mn+1.5La.

The SEM microstructures of the Mg-3Sn-1Mn and Mg-3Sn-1Mn-1.5La alloys in the backward-extrusion state are shown in Figure 7a,b. By comparison, it was found that more dense precipitates appeared on the matrix surface of the Mg-3Sn-1Mn-1.5La alloy. Combined with previous XRD analysis, it was found that the precipitates were mainly composed of a mixture of Mg_2Sn and MgSnLa. In addition, it can be observed from Figure 7c,d that the grains underwent apparent fragmentation during the extrusion deformation process, and the second phase was mainly distributed densely along the grain boundaries. From this, it can be seen that the distribution of the second phase was within the Mg crystal and at the grain boundaries. To further observe and analyze the structure and composition of the precipitates, Figure 8 shows the TEM microstructure and EDS energy spectrum analysis of the two alloys with the extrusion state. It can be seen from Figure 8a,c that spherical secondary precipitates were evenly distributed in the crystals of the two alloys. The selected area's electron diffraction pattern showed a single phase, and the EDS energy spectrum analysis was carried out at points A and B, respectively, as shown in Figure 8b,d. It can be determined as the Mg_2Sn phase and the Mg_2Sn phase in the crystal of the Mg-3Sn-1Mn-1.5La alloy was more refined and spheroidized than that of the Mg-3Sn-1Mn alloy. Zhao et al. [11,31,37] indicate that the TEM images and superimposed diffraction patterns of the matrix show that the Mg_2Sn phase was short-rod, lath-shaped, and spherical with hundreds of nanometers in thickness and diameter in the Mg-3Sn-1Mn alloy. However, the plate-shaped MgSnLa compound can be observed in the as-prepared Mg-3Sn-1Mn-1La alloy. Figure 9a–d shows the TEM structure morphology and EDS energy spectrum analysis of the Mg-3Sn-1Mn+1.5La alloy's grain boundary location. It can be seen that a large number of blocky and plate-shaped second phases were distributed at the grain boundary. The selected area's electron diffraction pattern showed that it was a symbiotic mixed phase. The EDS analysis results at the junction points A and B are shown in Figure 9b,d. According to the XRD phase analysis, the blocky or plate-shaped compounds were mainly a MgSnLa mixture composed of the La_5Sn_3, Mg_2Sn, and $Mg_{17}La_2$ phases. The spherical phase with smaller size (average diameter ≤ 40 nm) was the Mg_2Sn phase.

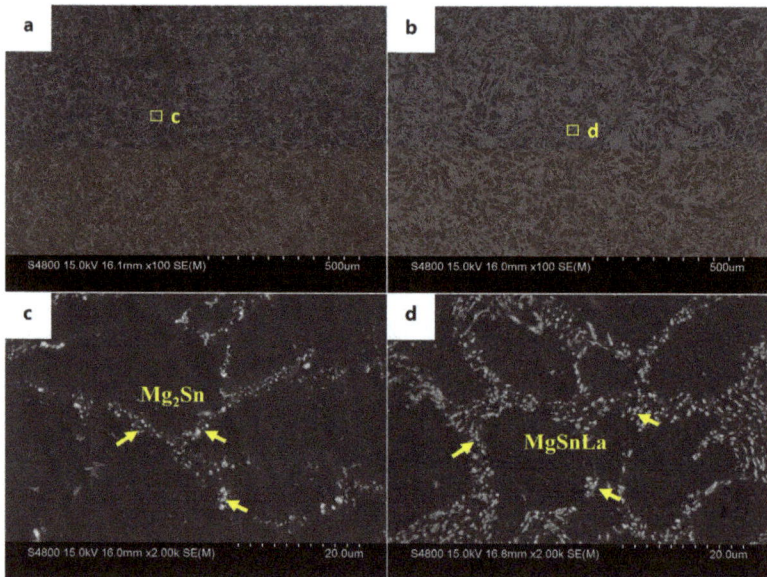

Figure 7. SEM microstructures of Mg-Sn-Mn-xLa backward-extruded alloys.

3.3. Tensile Mechanical Properties

Figure 10 shows the stress–strain response curve of Mg-3Sn-1Mn-xLa backward-extrusion alloy under a room-temperature tensile test. It can be seen from the figure that the stiffness of all specimens remained consistent during the initial linear elasticity stage. The ultimate tensile strength of the Mg-3Sn-1Mn alloy was 240 Mpa, and the elongation at tensile failure was 6.8%. After adding 0.5% La, the mechanical properties of the alloy underwent significant changes, with a substantial increase in ultimate tensile strength of 270 Mpa and an elongation of 9.2%. The effect of rare earth strengthening on mechanical properties was very obvious. As the La content continued to increase to 1%, the tensile strength of the alloy also further increased to 285 Mpa, with an elongation of 10.4% at this time. When the La content increased to 1.5%, the tensile strength of the alloy also reached 300 Mpa, and the elongation at this time was as high as 12.6%. When the La content continued to increase to 2%, the tensile strength of the alloy was 294 Mpa, and the elongation was 10.7%, indicating a specific decrease in performance. This decrease was probably caused by the excessive La, reducing the uniformity and refinement of the metal microstructure. From the above phenomena, we can conclude that adding La can effectively improve the mechanical properties of the Mg-3Sn-1Mn alloy. As the La content increases, the mechanical properties of the alloy also improve. However, when the La content reaches 2%, the alloy performance begins to decline, indicating that the addition of La should be controlled within a reasonable range. Compared with the Mg-3Sn-1Mn alloy without adding the La element, the tensile strength and elongation of the Mg-3Sn-1Mn-xLa alloy with a 1.5% content increased by 25% and 85%, respectively. This improvement has significance for the performance strengthening and optimization of magnesium alloys.

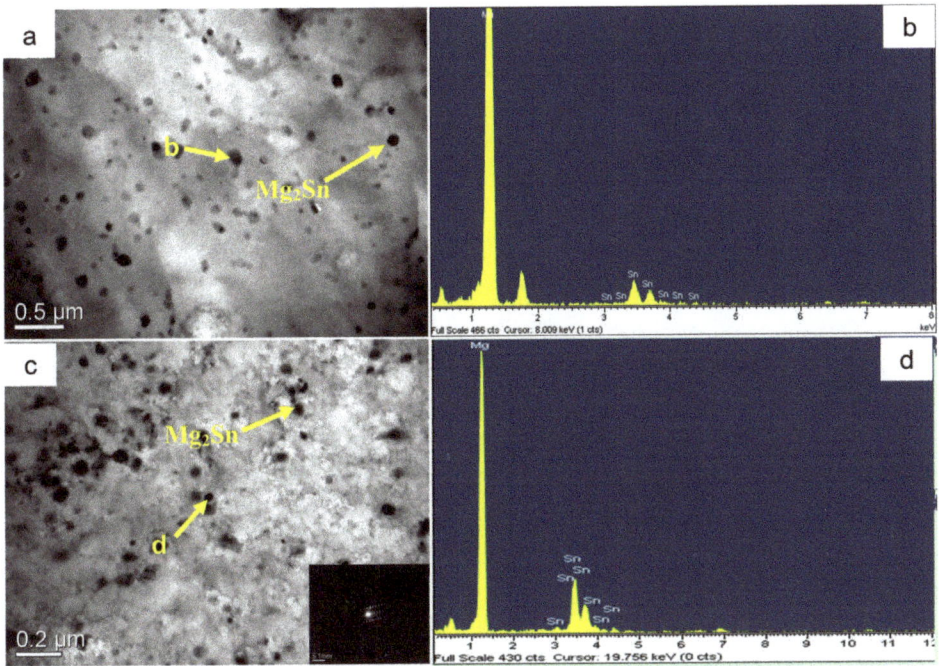

Figure 8. TEM microstructures of Mg-3Sn-1Mn-xLa and composition analysis. (**a**,**b**) Mg-3Sn-1Mn, (**c**,**d**) Mg-3Sn-1Mn-1.5La.

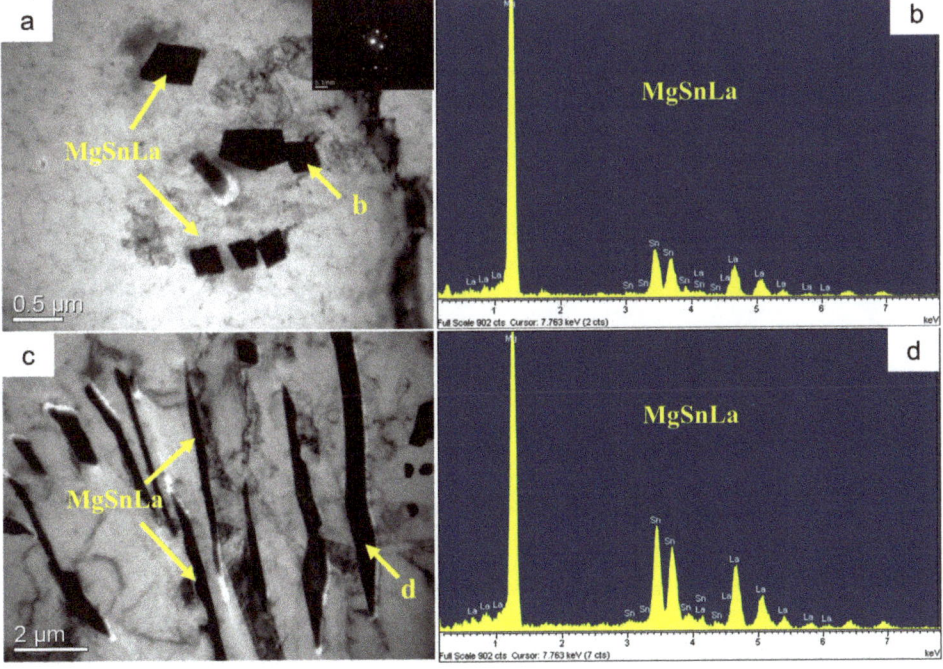

Figure 9. TEM microstructures of Mg-3Sn-1Mn-1.5La and composition analysis.

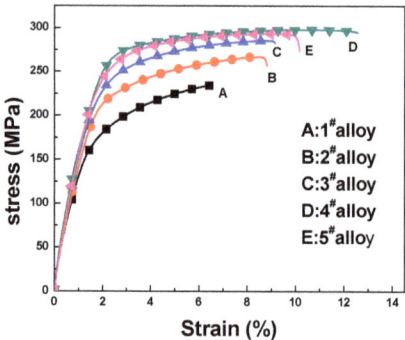

Figure 10. Stress–strain curves of Mg-3Sn-1Mn-xLa alloys with backward extruded.

Figure 11 shows the SEM fractography of the tensile-tested samples from backward-extruded parts at room temperature. It can be seen from Figure 11a that the fracture surface of the Mg-3Sn-1Mn alloy was mainly composed of rough tearing edges belonging to a typical brittle fracture. With the increase in La content, the fracture morphology of the alloy was significantly refined, and the grain distribution tended to be uniform, possessing the apparent refinement of an equiaxed grain fracture. At the same time, some small ductile dimples also appeared, with a mixed fracture feature of brittleness and toughness, as seen in Figure 11b. This was also the direct reason for the significant improvement in tensile strength and elongation with the addition of La seen in Figure 10. Continuing to increase the La content, the number of dimples on the alloy cross-section began to grow, and the tearing edges significantly decreased and became smaller. The fracture surface of the alloy began to appear as a cleavage layer, and the plastic characteristics further improved, with the typical characteristics of brittle fracture and cleavage fracture. The number of fracture dimples further increased, as shown in Figure 11c. When the La content continued to increase to 1.5%, the morphology of the fracture pits increased significantly, and the pits' depth also increased significantly. The cleavage phenomenon began to decrease; the tearing edges disappeared; and the slight dimples were tightly arranged. The grain refinement and homogenization phenomenon was more prominent, as shown in Figure 11d. However, when the La content reached 2.0 wt.%, the distribution of the fracture toughness pits began to coarsen (Figure 11e), which also confirms why the performance (Figure 10) began to decline.

3.4. Strengthening Mechanisms

In the Mg-3Sn-1Mn-xLa alloy, the Mn element is wholly dissolved in the α-Mg matrix and plays a solid solution strengthening role [33,42,43]. The size and distribution of the Mg_2Sn and Sn_3RE_5 phases have a significant impact on the mechanical properties of alloys [9,11,44,45]. Sn_3RE_5 is a cracked, rod-shaped phase formed by the combination of Sn and rare earth elements in Mg alloys. The Sn_3RE_5 phases exhibit superior high-temperature stability. The interface between the Mg_2Sn and Sn_3RE_5 phase and the α-Mg matrix is relatively stable, and it is difficult to nucleate and generate microcracks during deformation. Moreover, the formation of the Mg_2Sn phase can effectively hinder dislocation slip and improve the strength of the alloy. Figure 12 shows the TEM microstructure of the Mg-3Sn-1Mn-xLa alloy sample after tensile deformation. It can be seen from Figure 12a,b that Mg_2Sn in the Mg-3Sn-1Mn alloy had a pinning effect on dislocations, which can effectively hinder the slip of dislocations. When dislocations form and move between Mg_2Sn phases, due to the pinning effect of Mg_2Sn relative to dislocations, dislocations aggregate between the Mg_2Sn phase particles, which hinders the further diffusion of dislocations and improves the strength of the alloy. When the content of the La element reaches 1.5 wt.%, as seen in Figure 12c,d, the plate-shaped MgSnLa compounds formed near the grain boundary grow perpendicular to the grain boundary direction and extend into the grain, pinning the grain boundary to prevent the grain boundary sliding. At the same time, the plate-shaped MgSnLa compounds can also pin dislocations to prevent dislocation sliding. When the dislocations move to the plate-shaped compound, due to the blocking effect of the plate-shaped MgSnLa compounds on the dislocations, the dislocations accumulate around it. When dislocations move between the plate-shaped compounds, they are surrounded by plate-shaped MgSnLa compounds, which hinder their further movement, thus improving the strength of the Mg alloy. Furthermore, from Figure 12e, it is shown that the hindrance of Mg_2Sn relative dislocations within the crystal is evident. The interaction between dislocations and Mg_2Sn exhibits a bypass mechanism, resulting in Orowan strengthening [31]. Some studies have also shown that when dislocations encounter the Mg_2Sn phase, they bypass the phase to undergo Orowan strengthening, further improving the strength of the alloy [46,47]. As the content of the La element increases, the Mg_2Sn phase gradually refines and spheroidizes, and its distribution tends to be uniform. The pinning effect of Mg_2Sn relative to dislocations and its hindrance to dislocation movement is enhanced, strengthening the alloy. When the content of the La element reaches 2.0 wt.%, the plate-shaped MgSnLa compounds gradually increase, which makes the grain boundary coarser and more pronounced. The pinning effect of the plate-shaped compounds on the grain boundary is weakened, so the mechanical properties of the alloy are reduced.

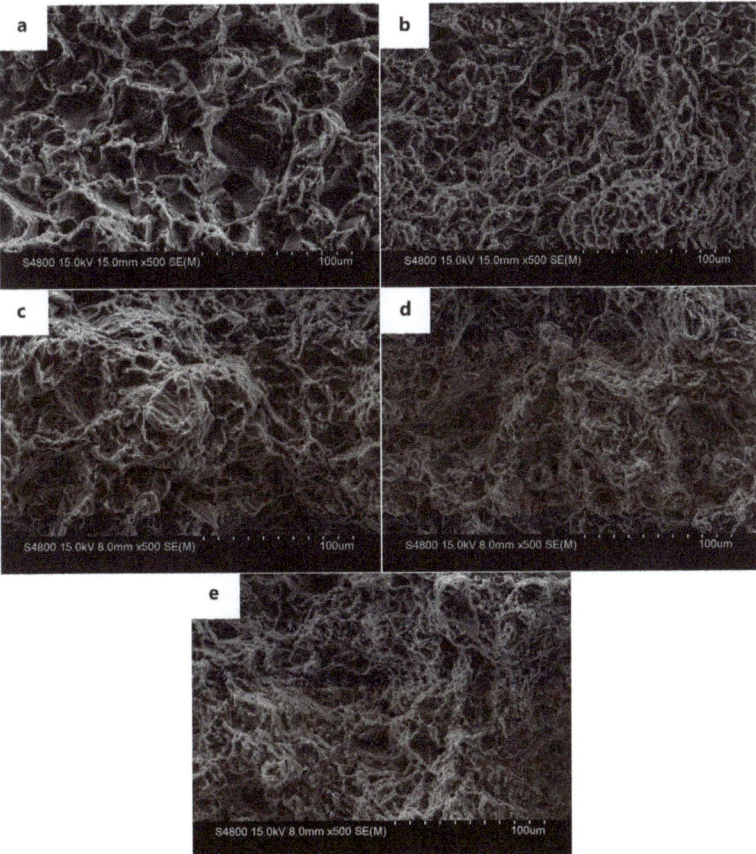

Figure 11. SEM micrographs of the fracture morphologies of the tested specimens: (**a**) Mg-3Sn-1Mn, (**b**) Mg-3Sn-1Mn+0.5La, (**c**) Mg-3Sn-1Mn+1.0La, (**d**) Mg-3Sn-1Mn+1.5La, (**e**) Mg-3Sn-1Mn+2.0La.

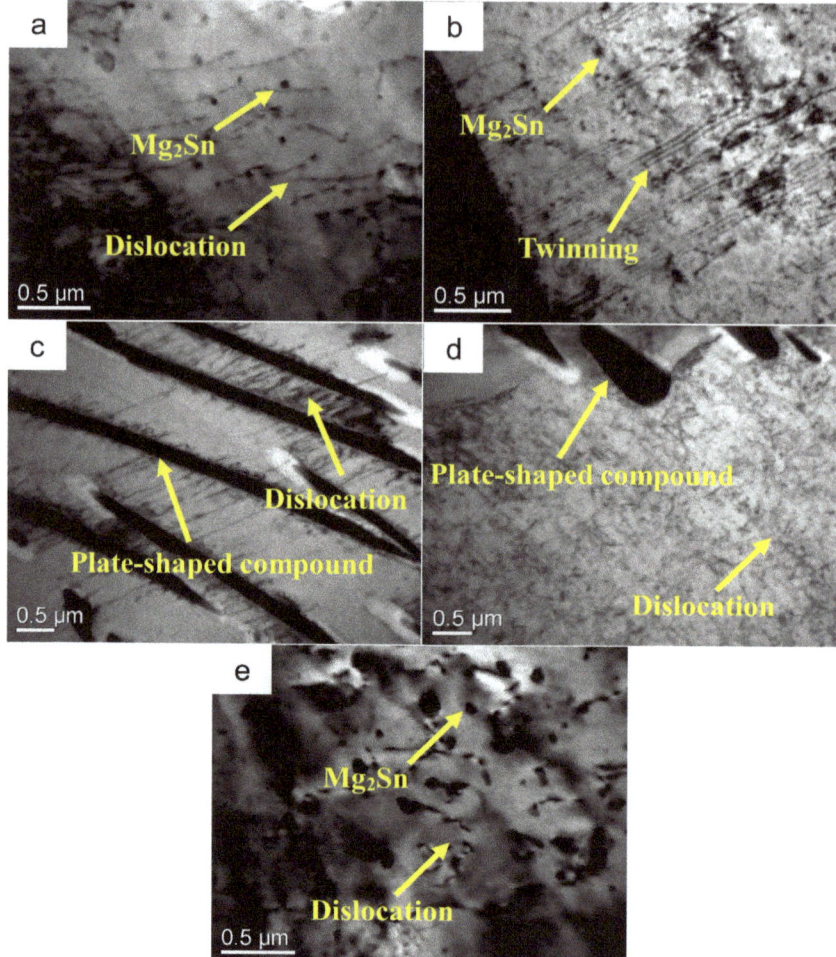

Figure 12. TEM images of backward-extrusion Mg alloy after tensile deformation (**a**,**b**) Mg-3Sn-1Mn, (**c**–**e**) Mg-3Sn-1Mn+1.5La.

4. Conclusions

This study investigated the microstructure evolution and mechanical properties of Mg-3Sn-1Mn-xLa (x = 0, 0.5, 1.5, 2.0 wt.%) alloys prepared using the backward-extrusion method. Based on the OM, XRD, SEM, and TEM observations and tensile properties, the main research findings are as follows:

1. In the cast and extruded Mg-3Sn-1Mn alloys, the main distribution was α-Mg grains and Mg_2Sn phases located at grain boundaries and within grains. Mg_2Sn phases can improve alloy properties by pinning dislocations and hindering dislocation slip.
2. With the addition of La, the plate-shaped MgSnLa compounds composed of $Mg_{17}La_2$, Mg_2Sn, and La_5Sn_3 phases began to be densely distributed along the α-Mg grain boundary, which can act as a pinning for grain boundaries and hinder dislocation slip. At the same time, the Mg_2Sn phases exhibited significant refinement and spheroidization.
3. With the increase in La, the mechanical properties of the extruded Mg-3Sn-1Mn-xLa alloy were significantly improved. When the La content reached 1.5%, the tensile strength at room temperature increased to 300 Mpa, and the elongation reached 12.6%,

i.e., 25% and 85% increases on the tensile strength (240 MPa) and elongation (6.8%) of the Mg-3Sn-1Mn alloy without La added, respectively. In addition, the grain size of rare earth magnesium alloy was significantly refined, and there were numerous small cleavage steps and dimples in the fracture surface of the sample. However, when the La content continued to increase to 2%, the alloy's properties began to show a downward trend, and there was a coarsening phenomenon in the distribution of the fracture surface.

Author Contributions: Conceptualization, Y.C.; Methodology, B.D.; Resources, X.Z.; Data curation, B.D.; Writing—original draft, X.Z.; Writing—review & editing, Y.C.; Supervision, X.Z.; Funding acquisition, Y.C. All authors have read and agreed to the published version of the manuscript.

Funding: This research was funded by the National Natural Science Foundation of China (52275165) and the Natural Science Foundation of Chongqing, China (CSTB2022NSCQ-MSX1290, cstc2021jcyj-msxmX0072, and cstc2019jscx-fxydX0036).

Institutional Review Board Statement: Not applicable.

Informed Consent Statement: Not applicable.

Data Availability Statement: The data presented in this study are available on request from the corresponding author.

Conflicts of Interest: The authors declare no conflict of interest.

References

1. Nazeer, F.; Long, J.; Yang, Z.; Li, C. Superplastic Deformation Behavior of Mg Alloys: A-Review. *J. Magnes. Alloy.* **2022**, *10*, 97–109. [CrossRef]
2. Liu, B.; Yang, J.; Zhang, X.; Yang, Q.; Zhang, J.; Li, X. Development and Application of Magnesium Alloy Parts for Automotive OEMs: A Review. *J. Magnes. Alloy.* **2023**, *11*, 15–47. [CrossRef]
3. Liu, B.Y.; Liu, F.; Yang, N.; Zhai, X.B.; Zhang, L.; Yang, Y.; Li, B.; Li, J.; Ma, E.; Nie, J.F.; et al. Large Plasticity in Magnesium Mediated by Pyramidal Dislocations. *Science* **2019**, *364*, 73–75. [CrossRef]
4. Wang, J.; Zhao, Z.; Bai, P.; Zhang, R.; Zhang, Z.; Wang, L.; Du, W.; Wang, F.; Huang, Z. Microstructure and Mechanical Properties of AZ31 Magnesium Alloy Prepared Using Wire Arc Additive Manufacturing. *J. Alloys Compd.* **2023**, *939*, 168665. [CrossRef]
5. Zheng, Y.; Zhang, Y.; Liu, Y.; Tian, Y.; Zheng, X.; Chen, L. Research Progress on Microstructure Evolution and Strengthening-Toughening Mechanism of Mg Alloys by Extrusion. *Materials* **2023**, *16*, 3791. [CrossRef]
6. Fan, R.; Wang, L.; Zhao, S.; Wang, L.; Guo, E. Strengthening of Mg Alloy with Multiple RE Elements with Ag and Zn Doping via Heat Treatment. *Materials* **2023**, *16*, 4155. [CrossRef]
7. Liu, C.Q.; Chen, H.W.; Liu, H.; Zhao, X.J.; Nie, J.F. Metastable Precipitate Phases in Mg–9.8 Wt%Sn Alloy. *Acta Mater.* **2018**, *144*, 590–600. [CrossRef]
8. Huang, X.; Huang, W. Irrational Crystallography of the ⟨1 1 2 0⟩Mg Mg$_2$Sn Precipitates in an Aged Mg-Sn-Mn Alloy. *Mater. Charact.* **2019**, *151*, 260–266. [CrossRef]
9. Deng, Y.; Sun, W.; Yang, Y.; Zhan, H.; Yan, K.; Zeng, G. Effects of Mg$_2$Sn Precipitation on the Age-Hardening and Deformation Behaviour of a Mg-Sn-Al-Zn Alloy. *Mater. Sci. Eng. A* **2023**, *867*, 144714. [CrossRef]
10. Liu, H.; Chen, Y.; Tang, Y.; Wei, S.; Niu, G. Tensile and Indentation Creep Behavior of Mg-5% Sn and Mg-5% Sn-2% Di Alloys. *Mater. Sci. Eng. A* **2007**, *464*, 124–128. [CrossRef]
11. Zhao, Z.Y.; Guan, R.G.; Shen, Y.F.; Bai, P.K. Grain Refinement Mechanism of Mg-3Sn-1Mn-1La Alloy during Accumulative Hot Rolling. *J. Mater. Sci. Technol.* **2021**, *91*, 251–261. [CrossRef]
12. Zhao, C.; Chen, X.; Pan, F.; Gao, S.; Zhao, D.; Liu, X. Effect of Sn Content on Strain Hardening Behavior of As-Extruded Mg-Sn Alloys. *Mater. Sci. Eng. A* **2018**, *713*, 244–252. [CrossRef]
13. Zhang, A.; Kang, R.; Wu, L.; Pan, H.; Xie, H.; Huang, Q.; Liu, Y.; Ai, Z.; Ma, L.; Ren, Y.; et al. A New Rare-Earth-Free Mg-Sn-Ca-Mn Wrought Alloy with Ultra-High Strength and Good Ductility. *Mater. Sci. Eng. A* **2019**, *754*, 269–274. [CrossRef]
14. Luo, Q.; Guo, Y.; Liu, B.; Feng, Y.; Zhang, J.; Li, Q.; Chou, K. Thermodynamics and Kinetics of Phase Transformation in Rare Earth–Magnesium Alloys: A Critical Review. *J. Mater. Sci. Technol.* **2020**, *44*, 171–190. [CrossRef]
15. Najafi, S.; Sheikhani, A.; Sabbaghian, M.; Nagy, P.; Fekete, K.; Gubicza, J. Modification of the Tensile Performance of an Extruded ZK60 Magnesium Alloy with the Addition of Rare Earth Elements. *Materials* **2023**, *16*, 2828. [CrossRef]
16. Du, Y.Z.; Qiao, X.G.; Zheng, M.Y.; Wu, K.; Xu, S.W. Development of High-Strength, Low-Cost Wrought Mg-2.5mass% Zn Alloy through Micro-Alloying with Ca and La. *Mater. Des.* **2015**, *85*, 549–557. [CrossRef]
17. Zhang, J.B.; Tong, L.B.; Xu, C.; Jiang, Z.H.; Cheng, L.R.; Kamado, S.; Zhang, H.J. Influence of Ca-Ce/La Synergistic Alloying on the Microstructure and Mechanical Properties of Extruded Mg–Zn Alloy. *Mater. Sci. Eng. A* **2017**, *708*, 11–20. [CrossRef]

18. Zengin, H.; Turen, Y. Effect of La Content and Extrusion Temperature on Microstructure, Texture and Mechanical Properties of Mg-Zn-Zr Magnesium Alloy. *Mater. Chem. Phys.* **2018**, *214*, 421–430. [CrossRef]
19. Luo, Q.; Zhai, C.; Gu, Q.; Zhu, W.; Li, Q. Experimental Study and Thermodynamic Evaluation of Mg–La–Zn System. *J. Alloys Compd.* **2020**, *814*, 152297. [CrossRef]
20. Zengin, H.; Ari, S.; Turan, M.E.; Hassel, A.W. Evolution of Microstructure, Mechanical Properties, and Corrosion Resistance of Mg–2.2Gd–2.2Zn–0.2Ca (Wt%)Alloy by Extrusion at Various Temperatures. *Materials* **2023**, *16*, 3075. [CrossRef]
21. Toda-Caraballo, I.; Galindo-Nava, E.I.; Rivera-Díaz-Del-Castillo, P.E.J. Understanding the Factors Influencing Yield Strength on Mg Alloys. *Acta Mater.* **2014**, *75*, 287–296. [CrossRef]
22. Son, H.W.; Hyun, S.K. Dislocation Characteristics and Dynamic Recrystallization in Hot Deformed AM30 and AZ31 Alloys. *J. Magnes. Alloy.* **2022**, *10*, 3495–3505. [CrossRef]
23. Mansoor, A.; Du, W.; Yu, Z.; Ding, N.; Fu, J.; Lou, F.; Liu, K.; Li, S. Effects of Grain Refinement and Precipitate Strengthening on Mechanical Properties of Double-Extruded Mg-12Gd-2Er-0.4Zr Alloy. *J. Alloys Compd.* **2022**, *898*, 162873. [CrossRef]
24. Wu, Q.; Yan, H.; Chen, J.; Xia, W.; Song, M.; Su, B. The Interactions between Dynamic Precipitates and Dynamic Recrystallization in Mg-5Zn-1Mn Alloys during Hot Compression. *Mater. Charact.* **2020**, *160*, 110131. [CrossRef]
25. Hoseini-Athar, M.M.; Mahmudi, R.; Prasath Babu, R.; Hedström, P. Effect of Zn Addition on Dynamic Recrystallization Behavior of Mg-2Gd Alloy during High-Temperature Deformation. *J. Alloys Compd.* **2019**, *806*, 1200–1206. [CrossRef]
26. Go, J.; Lee, J.U.; Yu, H.; Park, S.H. Influence of Bi Addition on Dynamic Recrystallization and Precipitation Behaviors during Hot Extrusion of Pure Mg. *J. Mater. Sci. Technol.* **2020**, *44*, 62–75. [CrossRef]
27. Meng, Y.; Yu, J.; Liu, K.; Yu, H.; Zhang, F.; Wu, Y.; Zhang, Z.; Luo, N.; Wang, H. The Evolution of Long-Period Stacking Ordered Phase and Its Effect on Dynamic Recrystallization in Mg-Gd-Y-Zn-Zr Alloy Processed by Repetitive Upsetting-Extrusion. *J. Alloys Compd.* **2020**, *828*, 154454. [CrossRef]
28. Chen, X.; Zhang, D.; Xu, J.; Feng, J.; Zhao, Y.; Jiang, B.; Pan, F. Improvement of Mechanical Properties of Hot Extruded and Age Treated Mg–Zn–Mn–Ca Alloy through Sn Addition. *J. Alloys Compd.* **2021**, *850*, 156711. [CrossRef]
29. Gu, X.J.; Cheng, W.L.; Cheng, S.M.; Liu, Y.H.; Wang, Z.F.; Yu, H.; Cui, Z.Q.; Wang, L.F.; Wang, H.X. Tailoring the Microstructure and Improving the Discharge Properties of Dilute Mg-Sn-Mn-Ca Alloy as Anode for Mg-Air Battery through Homogenization Prior to Extrusion. *J. Mater. Sci. Technol.* **2021**, *60*, 77–89. [CrossRef]
30. Guan, R.G.; Zhao, Z.Y.; Zhang, H.; Cui, T.; Lee, C.S. Microstructure and Properties of Mg-3Sn-1Mn (Wt%) Alloy Processed by a Novel Continuous Shearing and Rolling and Heat Treatment. *Mater. Sci. Eng. A* **2013**, *559*, 194–200. [CrossRef]
31. Guan, R.G.; Shen, Y.F.; Zhao, Z.Y.; Misra, R.D.K. Nanoscale Precipitates Strengthened Lanthanum-Bearing Mg-3Sn-1Mn Alloys through Continuous Rheo-Rolling. *Sci. Rep.* **2016**, *6*, 23154. [CrossRef]
32. Hua, Z.M.; Wang, C.; Wang, T.S.; Du, C.; Jin, S.B.; Sha, G.; Gao, Y.; Jia, H.L.; Zha, M.; Wang, H.Y. Large Hardening Response Mediated by Room-Temperature Dynamic Solute Clustering Behavior in a Dilute Mg-Zn-Ca-Sn-Mn Alloy. *Acta Mater.* **2022**, *240*, 118308. [CrossRef]
33. Hou, Y.; Qi, F.; Ye, Z.; Zhao, N.; Zhang, D.; Ouyang, X. Effects of Mn Addition on the Microstructure and Mechanical Properties of Mg–Sn Alloys. *Mater. Sci. Eng. A* **2020**, *774*, 138933. [CrossRef]
34. Zhong, L.; Wang, Y.; Dou, Y. On the Improved Tensile Strength and Ductility of Mg--Sn--Zn--Mn Alloy Processed by Aging Prior to Extrusion. *J. Magnes. Alloy.* **2019**, *7*, 637–647. [CrossRef]
35. Liao, H.; Kim, J.; Lv, J.; Jiang, B.; Chen, X.; Pan, F. Microstructure and Mechanical Properties with Various Pre-Treatment and Zn Content in Mg-Gd-Y-Zn Alloys. *J. Alloys Compd.* **2020**, *831*, 154873. [CrossRef]
36. Wei, X.; Jin, L.; Liu, C.; Wang, F.; Dong, S.; Dong, J. Effect of Pack-Forging on Microstructure and Properties of Mg-Gd-Y-Zn-Zr Alloy. *Mater. Sci. Eng. A* **2021**, *802*, 140674. [CrossRef]
37. Zhao, Z.; Bai, P.; Guan, R.; Murugadoss, V.; Liu, H.; Wang, X.; Guo, Z. Microstructural Evolution and Mechanical Strengthening Mechanism of Mg-3Sn-1Mn-1La Alloy after Heat Treatments. *Mater. Sci. Eng. A* **2018**, *734*, 200–209. [CrossRef]
38. Wang, J.H.; Zhao, Z.Y. First Principle Study of MgSnLa Compounds in Mg-3Sn-1Mn-1La Alloy Processed by Rheo-Rolling. *Materials* **2022**, *15*, 1361. [CrossRef]
39. *GB/T 16865-2013*; Test Pieces and Method for Tensile Test for Wrought Aluminium and Magnesium Alloys Products. Standardization Administration of the People's Republic of China: Beijing, China, 2013.
40. Zhao, X.; Li, S.; Zhang, Z.; Gao, P.; Kan, S.; Yan, F. Comparisons of Microstructure Homogeneity, Texture and Mechanical Properties of AZ80 Magnesium Alloy Fabricated by Annular Channel Angular Extrusion and Backward Extrusion. *J. Magnes. Alloy.* **2020**, *8*, 624–639. [CrossRef]
41. Gui, Y.; Ouyang, L.; Cui, Y.; Bian, H.; Li, Q.; Chiba, A. Grain Refinement and Weak-Textured Structures Based on the Dynamic Recrystallization of Mg–9.80Gd–3.78Y–1.12Sm–0.48Zr Alloy. *J. Magnes. Alloy.* **2021**, *9*, 456–466. [CrossRef]
42. Wan, D.; Wang, J.; Wang, G.; Chen, X.; Linlin; Feng, Z.; Yang, G. Effect of Mn on Damping Capacities, Mechanical Properties, and Corrosion Behaviour of High Damping Mg-3 Wt.%Ni Based Alloy. *Mater. Sci. Eng. A* **2008**, *494*, 139–142. [CrossRef]
43. Abdiyan, F.; Mahmudi, R.; Ghasemi, H.M. Effect of Mn Addition on the Microstructure, Mechanical Properties and Corrosion Resistance of a Biodegradable Mg–Gd–Zn Alloy. *Mater. Chem. Phys.* **2021**, *271*, 124878. [CrossRef]
44. Zhuo, X.; Zhao, L.; Gao, W.; Wu, Y.; Liu, H.; Zhang, P.; Hu, Z.; Jiang, J.; Ma, A. Recent Progress of Mg-Sn Based Alloys: The Relationship between Aging Response and Mechanical Performance. *J. Mater. Res. Technol.* **2022**, *21*, 186–211. [CrossRef]

45. Ding, Z.; Zhi, X.; Liu, B.; Hou, H.; Zhang, S.; Guo, W.; Chen, D.; Zhao, Y. Enhancement of Strength and Elastic Modulus of Mg-Gd-Y-Zn-Zr Alloy by Sn Addition. *Mater. Sci. Eng. A* **2022**, *854*, 143885. [CrossRef]
46. Sasaki, T.T.; Oh-ishi, K.; Ohkubo, T.; Hono, K. Effect of Double Aging and Microalloying on the Age Hardening Behavior of a Mg-Sn-Zn Alloy. *Mater. Sci. Eng. A* **2011**, *530*, 1–8. [CrossRef]
47. Fu, H.; Guo, J.; Wu, W.; Liu, B.; Peng, Q. High Pressure Aging Synthesis of a Hexagonal Mg_2Sn Strengthening Precipitate in Mg-Sn Alloys. *Mater. Lett.* **2015**, *157*, 172–175. [CrossRef]

Disclaimer/Publisher's Note: The statements, opinions and data contained in all publications are solely those of the individual author(s) and contributor(s) and not of MDPI and/or the editor(s). MDPI and/or the editor(s) disclaim responsibility for any injury to people or property resulting from any ideas, methods, instructions or products referred to in the content.

Article

Establishment of Constitutive Model and Analysis of Dynamic Recrystallization Kinetics of Mg-Bi-Ca Alloy during Hot Deformation

Qinghang Wang [1,2,*], Li Wang [1], Haowei Zhai [1], Yang Chen [3] and Shuai Chen [1]

1 School of Mechanical Engineering, Yangzhou University, Yangzhou 225127, China
2 School of Materials Science and Engineering, Hebei University of Technology, Tianjin 300130, China
3 School of Materials and Energy, Yunnan University, Kunming 650599, China
* Correspondence: wangqinghang@yzu.edu.cn; Tel.: +86-188-837-25047

Abstract: The flow behavior of the solution-treated Mg-3.2Bi-0.8Ca (BX31, wt.%) alloy was systematically investigated during hot compression at different deformation conditions. In the present study, the strain-related Arrhenius constitutive model and dynamic recrystallization (DRX) kinetic model were established, and the results showed that both two models had high predictability for the flow curves and the DRX behavior during hot compression. In addition, the hot processing maps were also made to confirm a suitable hot working range. Under the assistance of a hot processing map, the extrusion parameters were selected as 573 K and 0.5 mm/s. After extrusion, the as-extruded alloy exhibited a smooth surface, a fine DRX structure with weak off-basal texture and good strength–ductility synergy. The newly developed strong and ductile BX31 alloy will be helpful for enriching low-cost, high-performance wrought Mg alloy series for extensive applications in industries.

Keywords: Mg-Bi-Ca alloy; hot deformation; constitutive model; dynamic recrystallization kinetics; microstructure; mechanical properties

1. Introduction

Magnesium and its alloys, as one of the green materials, have shown wide application prospects in transportation, electronics and military industry due to its excellent characteristics, i.e., low density, high specific strength/stiffness, good vibration-reducing performance, etc. [1]. In recent years, Mg alloys have also made new breakthroughs in the field of hydrogen storage [2,3]. However, some bottleneck problems still exist, such as bad corrosion resistance, poor formability and strength–ductility trade of dilemma, to a large extent restricting commercial applications. In the past decade, some researchers have tried their best to overcome the imbalance of strength and ductility using rare-earth (RE) elements [4,5] and severe plastic deformation techniques [6]. Nevertheless, the high cost and the complex processing vastly limit the applications of Mg alloys. Thus, the development of low RE-containing, even RE-free Mg alloys, is an imminent requirement.

Up to now, some new Mg alloys, such as Mg-Sm- [7,8], Mg-Ca- [9,10], Mg-Al- [11], Mg-Sn- [12], Mg-Mn- [13,14] and Mg-Zn-based [15] alloys, have been fabricated successfully to achieve this goal. In order to further enrich the existing low-cost, high-performance Mg alloy systems, Mg-Bi-based series alloys have been exploited in recent years, and they have great potential in developing RE-free materials with outstanding comprehensive mechanical properties. Al, Zn, Ca, Mn and Sn elements have been attempted to add into Mg-Bi-based alloys to ameliorate their mechanical properties [16–23]. Among them, interestingly, wrought Mg-Bi-Ca series alloys exhibit unique mechanical properties, which are dependent on hot deformation parameters. For instance, Meng et al. [17–19] pointed out that Ca addition could trigger the texture change in as-extruded Mg-Bi-Ca alloys. As extrusion temperature or die-exit speed increased, an off-basal texture feature formed, inducing a high tensile ductility (~40%) in Mg-1.3Bi-0.9Ca (wt.%) alloy [17,18]. In contrast, under the conditions of a low extrusion temperature of 280 °C and die-exit speed of 4 mm/s, a strong basal texture occurred in Mg-1.5Bi-0.8Ca (wt.%) alloy leading to a high tensile yield strength (~394 MPa) [19]. In order to obtain the strength–ductility synergic Mg-Bi-Ca series alloys, a two-step deformed Mg-1.3Bi-0.7Ca (wt.%) alloy, with a tensile yield strength of ~351 MPa and elongation-to-failure of ~13.2%, was successfully prepared by extrusion and subsequent caliber rolling [20]. In summary, such a new highly strong and ductile RE-free Mg-Bi-Ca alloy will be helpful for enriching low-cost and high-performance wrought Mg alloy series to achieve extensive applications in industries.

However, until now, the flow behavior, constitutive model, dynamic recrystallization (DRX) kinetic model and hot processing map of Mg-Bi-Ca series alloys during hot deformation are rarely reported systematically. It is important to understand the hot flow behavior of Mg-Bi-Ca series alloys to obtain the desired microstructure and to achieve excellent mechanical properties. The Arrhenius-type constitutive model is commonly used to describe the hot flow behavior, predicting the hot-deformed microstructure and optimizing the hot processing parameters of Mg alloys [24,25]. In Mg alloys, the chemical composition is a key factor in influencing the DRX kinetic. When the added alloying elements are soluble in a substrate, non-basal slip may be activated easily during hot deformation since these elements (i.e., RE, Ca and Li) obviously change the stacking fault energy for non-basal slip reducing the ratio of critical resolved shear stresses, (CRSSs) between non-basal and basal slips [26–28]. Within grains, the activation of massive slip systems accelerates the dislocation rearrangement to form low-angle grain boundaries (LAGBs), which provide a precursor for the generation of large-angle grain boundaries (HAGBs). For example, Li et al. [29] pointed out that Ce addition into AZ80 alloy presented a lower activation energy than that in without Ce. In addition to non-basal slip, contraction twinning and/or double twinning also could occur in the early stage of hot deformation in Y-containing Mg-Sn-Zn alloy, as reported by Wang et al. [30]. These twins offer the DRX nucleation sites, also promoting the DRX process. On the other hand, as alloying elements are presented as compounds, two cases exist. In general, large-sized phases (>1 μm) play a role in increasing the DRX nucleation sites, known as particles stimulate nucleation (PSN). While the phase size is below 1 μm, they tend to delay the motion of grain boundaries inhibiting the DRX process. However, on the DRX mechanism of Mg-Bi-Ca series alloys, it is still unknown.

Therefore, in order to investigate the hot deformation behavior of Mg-Bi-Ca series alloys, in this work, taking a new RE-free Mg-Bi-Ca alloy as an example, hot compressive tests were performed at different conditions. The Arrhenius constitutive model and the DRX kinetics of this alloy were established. Based on the data of hot compressive flow curves, the suitable hot processing range was measured. Finally, the extrusion parameter was selected to gain the fine complete DRX structure with weak off-texture and obtain the strength–ductility balance in an as-extruded Mg-Bi-Ca alloy.

2. Materials and Methods

Commercial pure Mg ingot (≥99.99%), Mg-10Bi (wt.%) and Mg-25Ca (wt.%) master alloys were melted at 720 °C in an electric resistance furnace (Shanghai Yuzhi Technology Co., Ltd, China) under a protective atmosphere (SF_6:CO_2 = 1:99). After a series of sitting and slagging, the melt was poured into a steel mold with Φ 80 (in diameter) × 200 (in height) mm preheated to 350 °C. The chemical composition of the as-cast alloy was detected by an X-ray fluorescence spectrometer (XRF, LAB CENTER XRF-1800) (Shimadzu, Kyoto, Japan), and the real chemical composition is 96.0 wt.% Mg, 3.2 wt.% Bi, and 0.8 wt.% Ca. It can be labeled as BX31. Before hot compression, the solid solution of the as-cast BX31 alloy was treated at 773 K for 24 h. Subsequently, hot compression tests for samples with a size of Φ 8 × 12 mm (diameter × height) were operated on a Gleeble-3500 thermomechanical simulator (DSI North America Corp, Marlton, Evesham, United States) at different conditions (temperatures: 573, 623 and 673 K; strain rates: 0.01, 0.1 and 1 s^{-1}). In addition, the solution-treated BX31 alloy was also extruded into the bar with Φ 22 mm in diameter at 573 K using an extrusion ratio of 21:1 and a die-exit speed of 0.5 mm/s by XJ-500 horizontal extruder (WuxiYuanchang Machine Manufacture Co., Wuxi, China). The room-temperature (RT) tensile properties of the as-extruded sample with 25 mm in gauge length and Φ 5 mm in gauge diameter were machined from the as-extruded bar along extrusion direction (ED) and measured using CMT6305-300 kN universal tensile testing machine (MTS Systems Co Ltd, Shanghai, China) at a strain rate of 1×10^{-3} s^{-1}.

The microstructure could be observed by scanning electron microscopy (SEM, Gemini SEM 300) (Carl Zeiss, Oberkochen, Germany) equipped with an energy dispersive spectrometer (EDS) (Carl Zeiss, Oberkochen, Germany) and electron backscattered diffraction (EBSD, JEOL JSM-7800F) device (Japan Electronics Corporation, Tokyo, Japan). The phase constitutions were identified by X-ray diffraction (XRD, Rigaku D/Max 2500) (Bruker AXS, Karlsruhe, Germany). The preparation of EBSD samples was composed of grinding, washing, blow-drying and electro-polishing at 20 V and 0.03 A for 90 s at 298 K using AC2. The scanning step size was set as 0.5 µm. All EBSD data were analyzed using ATEX software v2.01.3 (ATEX, Metz, France).

3. Results and Discussion

3.1. Microstructure of the Solution-Treated BX31 Alloy

Figure 1a shows the SEM image of the solution-treated BX31 alloy with a few bright granular-shaped second phases. The average phase size is about 5 µm. Based on the SEM-EDS mapping scanning results, as shown in Figure 1b–d, these granular-shaped phases mainly consist of Mg, Bi and Ca elements possibly deemed as Mg-Bi-Ca ternary phases. Moreover, by combining the SEM-EDS point scanning result from phase 1 marked by the red arrow in Figure 1a, these ternary phases can be regarded as Mg_2Bi_2Ca ones, as displayed in Figure 1e. The reports of Remennik et al. [31], Meng et al. [17–20] and Liu et al. [32] are consistent with this result. In addition, Mg_3Bi_2 phases, even Mg_2Ca ones, also could be found in wrought Mg-Bi-Ca series alloys [17–20]. However, the mentioned these two kinds of phases are almost undetectable in the solution-treated BX31 alloy. We argue that these dynamically precipitated fine Mg_3Bi_2 and Mg_2Ca particles are the products of a combined effect from deformation force and temperature. On the other hand, the thermal stability of Mg_2Bi_2Ca phases is superior to Mg_3Bi_2 and Mg_2Ca ones, based on the thermodynamics of phase formation [32]. Therefore, in this work, the main phase constitution in the solution-treated BX31 alloy is Mg_2Bi_2Ca. In order to further verify this statement, an XRD test was carried out, and the result is shown in Figure 2. We can see that two kinds of characteristic diffraction peaks of phases exist: α-Mg and Mg_2Bi_2Ca, which is in accordance with the SEM observation.

3.2. Flow Behavior during Hot Compression

Figure 3 shows the compressive flow curves of the solution-treated BX31 alloy at different deformation conditions. We can see the flow curves contain three stages: (1) Rapid ascent stage. Dislocations accumulate sharply, giving rise to remarkable work hardening; (2) Slow descent stage. The flow stress reaches the peak as the strain increases, and subsequently, dislocations annihilate, giving rise to the decline of the flow stress. This is mainly attributed to the DRX triggering the dynamic softening [24,25,33]; (3) Steady-state flow stage. In this stage, the work hardening and the dynamic softening reach a balance, thereby making the flow stress tend to accord. Figure 4 shows the peak stresses and the corresponding peak strains at different deformation conditions. There exists an obvious tendency for the peak stress increment by decreasing the temperature and increasing the strain rate (see Figure 4a), but the peak strain tends to be decreased (see Figure 4b). This phenomenon is closely associated with the acceleration of the DRX process.

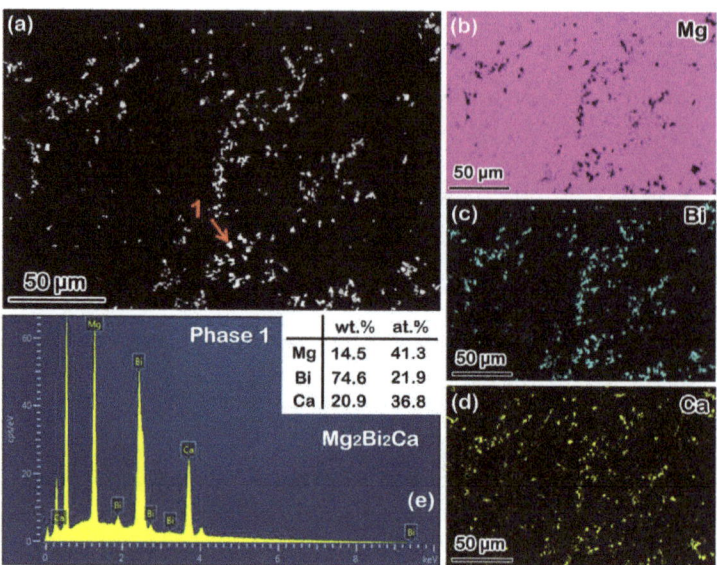

Figure 1. (a) SEM image; (b–d) SEM-EDS mapping scanning results from (a); (e) SEM-EDS point scanning result from phase 1 marked by red arrow in (a).

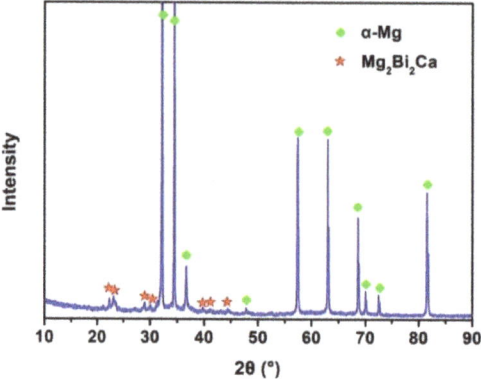

Figure 2. XRD result of the solution-treated BX31 alloy.

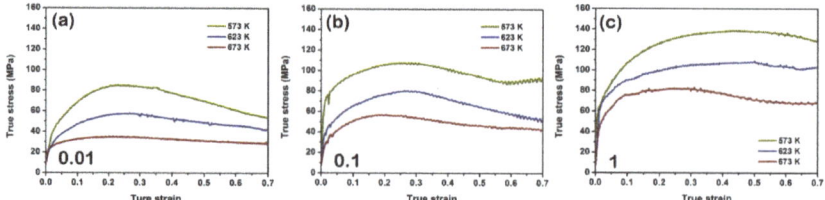

Figure 3. True compressive stress–strain curves at strain rates: (a) $0.01\ \text{s}^{-1}$; (b) $0.1\ \text{s}^{-1}$; (c) $1\ \text{s}^{-1}$.

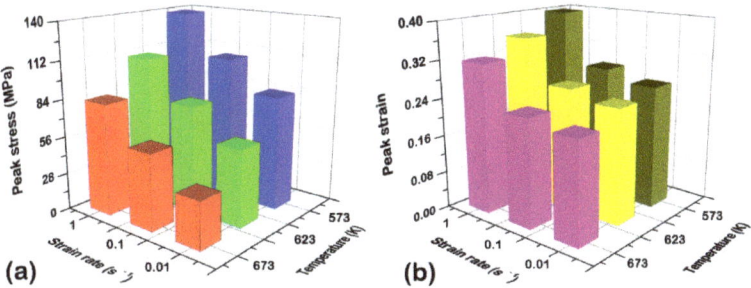

Figure 4. (a) Peak stresses and (b) peak strains at different deformation conditions. Note that the color difference is to better distinguish the size of the value.

3.3. Constitutive Model during Hot Compression

The Arrhenius constitutive models are given [34]:

$$\dot{\varepsilon} = A_1 \sigma^{n_1} exp(-\frac{Q}{RT}) \tag{1}$$

$$\dot{\varepsilon} = A_2 exp(\beta\sigma) exp(-\frac{Q}{RT}) \tag{2}$$

$$\dot{\varepsilon} = A[sinh(\alpha\sigma)]^n exp(-\frac{Q}{RT}) \tag{3}$$

where $\dot{\varepsilon}$ is the strain rate; σ is the flow stress; T is the temperature; A_1, n_1, A_2, β, A, α and n are material constants; Q is the activation energy; and R is the mole gas constant (8.314 J/mol K). The n_1, β and α can be described as follows [35]:

$$\alpha = \frac{\beta}{n_1} \tag{4}$$

Zener–Hollomon (Z) parameter is introduced to construct the relationship among the Z, $\dot{\varepsilon}$ and T. It can be shown as follows [36]:

$$Z = \dot{\varepsilon} exp(\frac{Q}{RT}) \tag{5}$$

For Equation (3) with Equation (5) together, the σ as a function of the Z is expressed as follows [37]:

$$\sigma = \frac{1}{\alpha} ln\left\{ \left(\frac{Z}{A}\right)^{\frac{1}{n}} + \left[\left(\frac{Z}{A}\right)^{\frac{2}{n}} + 1\right]^{\frac{1}{2}} \right\} \tag{6}$$

Under the peak stress condition, these material constants of α, Q, n and A are calculated by Equations (7)–(11).

Both sides of Equations (1) and (2) are taken natural logarithms as follows:

$$ln\dot{\varepsilon} = n_1 ln\sigma + lnA_1 - \frac{Q}{RT} \tag{7}$$

$$ln\dot{\varepsilon} = \beta\sigma + lnA_2 - \frac{Q}{RT} \tag{8}$$

Figure 5a,b show the linear relationships of $ln\dot{\varepsilon} - ln\sigma$ and $ln\dot{\varepsilon} - \sigma$, respectively. The slopes of these two functions are considered the average n_1 and β values (~7.317 and ~0.089, respectively). According to Equation (4), the α is estimated by ~0.012.

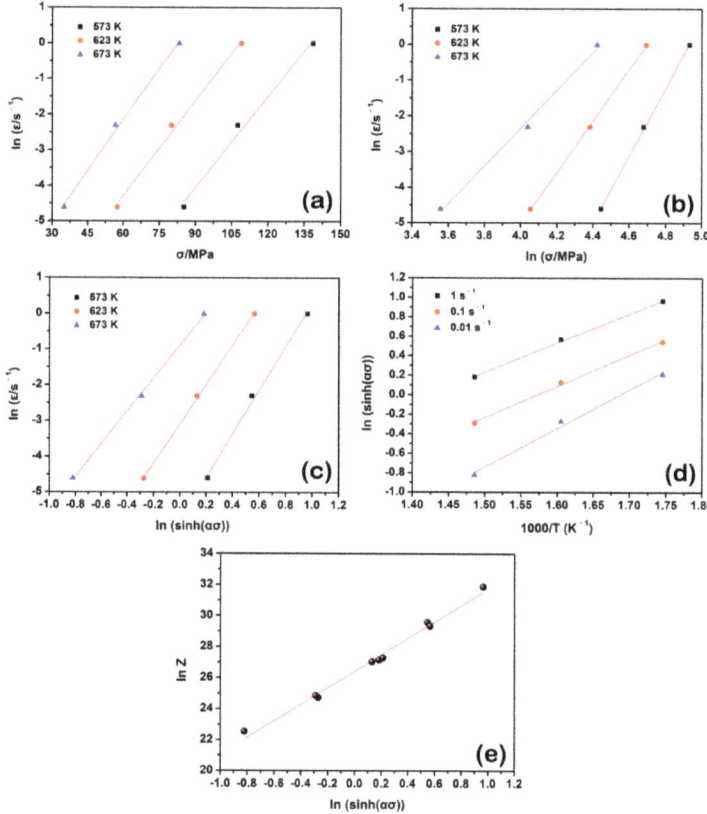

Figure 5. Linear relations of (**a**) $ln\dot{\varepsilon} - ln\sigma_p$; (**b**) $ln\dot{\varepsilon} - \sigma_p$; (**c**) $ln\dot{\varepsilon} - ln[sin\,h(\alpha\sigma_p)]$; (**d**) $ln[sin\,h(\alpha\sigma)] - 1/T$ and (**e**) $lnZ - ln[sin\,h(\alpha\sigma_p)]$ under the peak stress condition of the solution-treated BX31 alloy.

Both sides of Equation (3) are taken natural logarithms as follows:

$$ln\dot{\varepsilon} = n ln[sin\,h(\alpha\sigma)] + lnA - \frac{Q}{RT} \qquad (9)$$

Figure 5c plots the functional relation between $ln\dot{\varepsilon}$ and $ln[sinh(\alpha\sigma)]$, and the slope can be regarded as the average n value (~5.390).

Both sides of Equation (9) are taken partial derivatives as follows:

$$Q = R\left(\frac{\partial ln\dot{\varepsilon}}{\partial ln[sin\,h(\alpha\sigma)]}\bigg|T\right)\left(\frac{\partial ln[sin\,h(\alpha\sigma)]}{\partial(1/T)}\bigg|\dot{\varepsilon}\right) = RnS \qquad (10)$$

Figure 5d shows the functional relation of $ln[sin\,h(\alpha\sigma)] - 1/T$, and the average S value is the slope of this linear relationship, ~3.390. Therefore, the Q value at the peak stress condition is about 151.9 kJ/mol, slightly larger than that of pure Mg (~135 kJ/mol [38]). Fine precipitates could effectively inhibit dislocation movement during deformation, which led to an increase in the activation energy during deformation [39]. According to the reports from the literature [17–20], fine Mg_3Bi_2 and Mg_2Ca particles may be dynamically precipitated from the Mg matrix in Mg-Bi-Ca series alloys, and a part of them was found to distribute at grain boundaries. Such a grain boundary pinning effect further enhances the resistance of grain boundary motion, giving rise to the increment of the activation energy during hot compression. In addition, recently, the Q value has been successfully related to the atomistic mechanisms for different engineering materials. Savaedi et al. [40] pointed out that the average Q value was close to the weighted self-diffusion activation energy for each element in CoCrFeMnNi alloy. Relevant research was also presented in the report of Jeong et al. [41]. This kind of analysis, considering the atomistic mechanisms, needs more attention in future work.

Combining Equation (3) with Equation (5), the following relationship is shown:

$$lnZ = nln[sinh(\alpha\sigma)] + lnA \tag{11}$$

The values of Z can be calculated at different deformation conditions by taking into the Q value. Figure 5e matches the linear function of $lnZ - ln[sinh(\alpha\sigma)]$, and the lnA is ~26.430 as the intercept of this function. As mentioned above, the Arrhenius constitutive model at the peak stress condition can be established as follows:

$$\dot{\varepsilon} = 3.008 \times 10^{11} [sinh(0.012\sigma)]^{5.390} exp(-\frac{1.827 \times 10^4}{T}) \tag{12}$$

Under the case of the strain-related Arrhenius constitutive model, the material constants of α, Q, n and A are closely associated with the strain. By using 6-order polynomial equations expresses their relations as follows:

$$\alpha(\varepsilon) = \alpha_6\varepsilon^6 + \alpha_5\varepsilon^5 + \alpha_4\varepsilon^4 + \alpha_3\varepsilon^3 + \alpha_2\varepsilon^2 + \alpha_1\varepsilon + \alpha_0 \tag{13}$$

$$Q(\varepsilon) = Q_6\varepsilon^6 + Q_5\varepsilon^5 + Q_4\varepsilon^4 + Q_3\varepsilon^3 + Q_2\varepsilon^2 + Q_1\varepsilon + Q_0 \tag{14}$$

$$n(\varepsilon) = n_6\varepsilon^6 + n_5\varepsilon^5 + n_4\varepsilon^4 + n_3\varepsilon^3 + n_2\varepsilon^2 + n_1\varepsilon + n_0 \tag{15}$$

$$A(\varepsilon) = exp[lnA(\varepsilon)] = exp(A_6\varepsilon^6 + A_5\varepsilon^5 + A_4\varepsilon^4 + A_3\varepsilon^3 + A_2\varepsilon^2 + A_1\varepsilon + A_0) \tag{16}$$

where ε is the strain. The coefficients $\alpha_0 - \alpha_6$, $Q_0 - Q_6$, $n_0 - n_6$ and $lnA_0 - lnA_6$ are listed in Table 1. Figure 6 shows these material parameters as functions of the strain. Based on the strain-related Arrhenius constitutive model, the predicted flow stress values at different strains for each deformation condition can be calculated, and Figure 7 compares the experiment flow stresses and the predicted values at different deformation conditions. It can be seen that the strain-related Arrhenius constitutive model is suitable for the flow curves of the solution-treated BX31 alloy during hot compression. The predictability of this model can be evaluated by the correlation coefficient (R) and the average absolute relative error ($AARE$), and the relative error (δ) as follows [37,42]:

$$R = \frac{\sum_{i=1}^{n}(E_I - \overline{E})(P_i - \overline{P})}{\sqrt{\sum_{i=1}^{n}(E_i - \overline{E})^2 \sum_{i=1}^{n}(P_i - \overline{P})^2}} \tag{17}$$

$$AARE(\%) = \frac{1}{n}\sum_{i=1}^{n}\left|\frac{E_i - P_i}{E_i}\right| \times 100\% \tag{18}$$

$$\delta(\%) = \left(\frac{E_i - P_i}{E_i}\right) \times 100\% \tag{19}$$

where n is the size of flow stress set. E_i and P_i are the values of experiment flow stress and predicted one, respectively. $\overline{E}$ and $\overline{P}$ are the mean values of E_i and P_i, respectively. The calculated result shows the R is ~0.9919 and the $AARE$ is ~4.28%, as shown in Figure 8a. Additionally, the δ is deviated at a low range of ~−1.147%, as shown in Figure 8b. It further suggests that this model works well.

Table 1. Coefficients in the polynomial fitting of α, Q, n and lnA.

α	Q	n	lnA
$\alpha_0 = 0.0343$	$Q_0 = 138.6302$	$n_0 = 5.6236$	$lnA_0 = 22.9529$
$\alpha_1 = -0.5172$	$Q_1 = -1049.6868$	$n_1 = -22.5556$	$lnA_1 = -181.8720$
$\alpha_2 = 5.1075$	$Q_2 = 18,785.7115$	$n_2 = 432.5720$	$lnA_2 = 3422.1192$
$\alpha_3 = -25.5741$	$Q_3 = -110,944.6683$	$n_3 = -2667.7302$	$lnA_3 = -20,427.6000$
$\alpha_4 = 67.i.e60$	$Q_4 = 303,759.5483$	$n_4 = 7324.5991$	$lnA_4 = 56,149.0301$
$\alpha_5 = -88.9583$	$Q_5 = -398,898.1624$	$n_5 = -9495.3504$	$lnA_5 = -73,886.5001$
$\alpha_6 = 46.1098$	$Q_6 = 203,225.5371$	$n_6 = 4756.1032$	$lnA_6 = 37,684.4203$

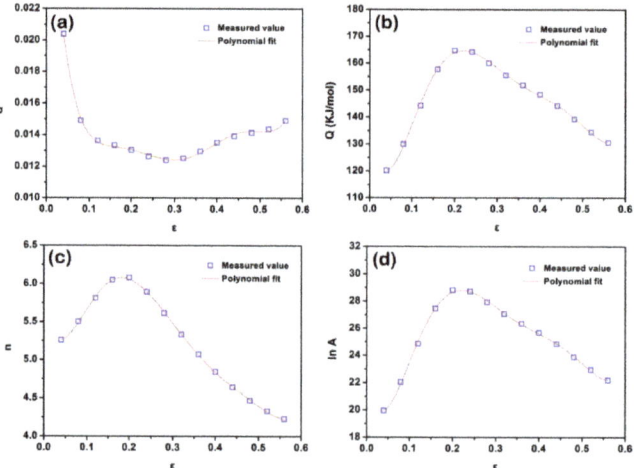

Figure 6. (**a**) α, (**b**) Q, (**c**) n and (**d**) lnA as functions of the true strain ε.

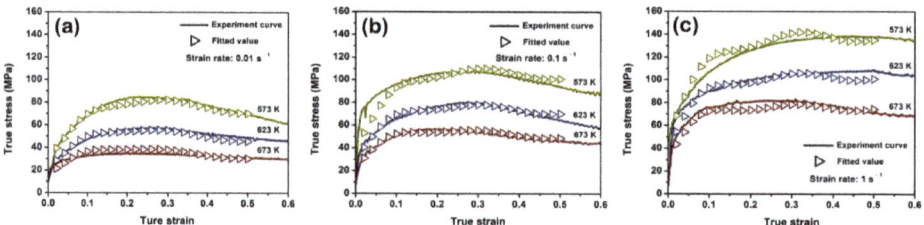

Figure 7. Comparison of experiment flow curves and predicted results of the strain-related Arrhenius constitutive model at different strain rates: (**a**) 0.01 s^{-1}; (**b**) 0.1 s^{-1}; (**c**) 1 s^{-1}.

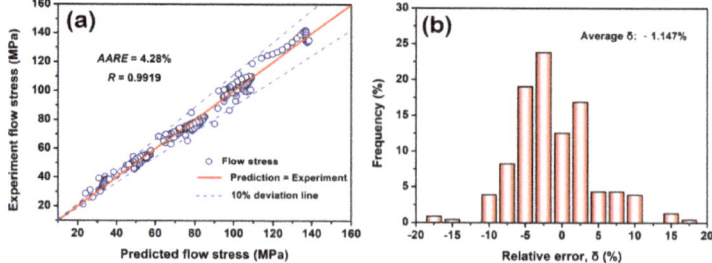

Figure 8. Correlation (**a**) and the relative error (**b**) between experiment and predicted flow stresses by the strain-related Arrhenius constitutive model.

3.4. DRX Kinetics during Hot Compression

In order to describe the DRX kinetics model, in general, the following equation can be used and expressed [43,44]:

$$X_{DRX} = 1 - exp\left[-k_D\left(\frac{\varepsilon - \varepsilon_c}{\varepsilon_{0.5}}\right)^{n_D}\right] \quad (20)$$

where X_{DRX} is the DRX area fraction, ε_c is the DRX critical strain, $\varepsilon_{0.5}$ is the strain for the formation of 50% DRX, k_D and n_D are the material constants. ε_c and $\varepsilon_{0.5}$ can be as a function of the Z parameter given as follows [44]:

$$\varepsilon_c = B_1 Z^{m_1} \quad (21)$$

$$\varepsilon_{0.5} = B_2 Z^{m_2} \quad (22)$$

where B_1, B_2, m_1 and m_2 are material constants. The corresponding ε_c and $\varepsilon_{0.5}$ values are obtained by calculating the σ_c and $\sigma_{0.5}$ ones that represent the stress values, and the corresponding initial DRX and 50% DRX occur, respectively. According to the θ-σ relationship (θ represents the work hardening rate), the σ_c can be gained. By taking a special case (573 K and 0.01 s^{-1}) as an example in Figure 9a, the key flow stresses, i.e., σ_c, peak stress (σ_p), saturated stress (σ_{sat}) and steady stress (σ_{ss}), can be measured. In addition, the $\sigma_{0.5}$ can be calculated as follows [44]:

$$X_{DRX}(50\%) = \frac{\sigma_p - \sigma_{0.5}}{\sigma_p - \sigma_{ss}} \quad (23)$$

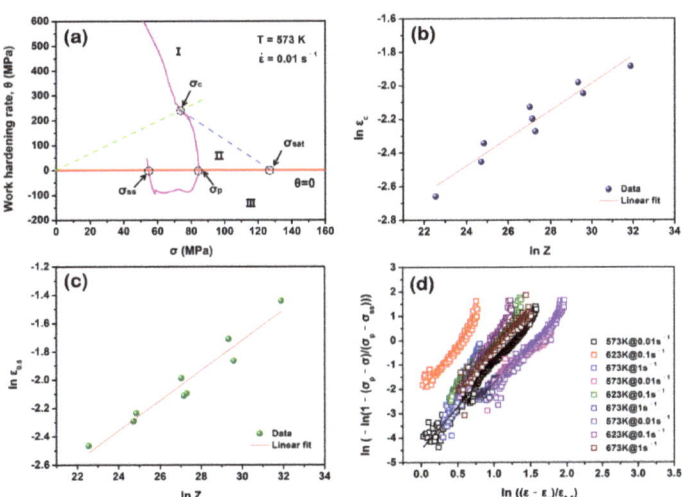

Figure 9. (a) θ-σ curve at 573 K and 0.01 s^{-1}. Note that it is divided into three stages marked by I, II and III, respectively; (b) relationship between $ln\ \varepsilon_c$ and $ln\ Z$; (c) relationship between $ln\ \varepsilon_{0.5}$ and $ln\ Z$; (d) $ln\ (-ln\ ((1 - (\sigma_p - \sigma)/(\sigma_p - \sigma_{ss}))))$ as a function of $ln\ ((\varepsilon - \varepsilon_c)/\varepsilon_{0.5})$.

According to the calculation o θ-σ curve and Equation (23), Table 2 lists the σ_c, $\sigma_{0.5}$, ε_c and $\varepsilon_{0.5}$ values at different deformation conditions. Figure 9b,c match the linear functions from $ln\varepsilon_c - lnZ$ and, $ln\varepsilon_{0.5} - lnZ$ respectively. Finally, the ε_c and $\varepsilon_{0.5}$ can be expressed:

$$\varepsilon_c = 0.0116 Z^{0.0823} \quad (24)$$

$$\varepsilon_{0.5} = 0.0073 Z^{0.1067} \quad (25)$$

Table 2. σ_c, $\sigma_{0.5}$, ε_c and $\varepsilon_{0.5}$ values at different deformation conditions.

	Strain Rate (s^{-1})	Deformation Temperature (K)		
		573	623	673
σ_c (MPa)	0.01	67.5	42.9	31.9
	0.1	93.9	56.4	50.3
	1	125.8	78.4	72.8
$\sigma_{0.5}$ (MPa)	0.01	69.1	47.2	32.0
	0.1	102.8	67.5	51.2
	1	130.7	88.8	74.8
ε_c	0.01	0.103	0.086	0.070
	0.1	0.129	0.119	0.096
	1	0.152	0.138	0.111
$\varepsilon_{0.5}$	0.01	0.123	0.101	0.085
	0.1	0.155	0.137	0.107
	1	0.237	0.181	0.121

Figure 9d shows the linear relationship between $ln\left(-ln\left(1 - \frac{\sigma_p - \sigma}{\sigma_p - \sigma_{ss}}\right)\right)$ and $ln\left(\frac{\varepsilon - \varepsilon_c}{\varepsilon_{0.5}}\right)$ at different deformation conditions, and the average k_D (~0.015) and n_D (~3.781) can be determined by the intercept and slope of this function, respectively. Therefore, the DRX kinetics model of the solution-treated BX31 alloy can be expressed as follows:

$$X_{DRX} = 1 - exp\left[-0.015\left(\frac{\varepsilon - 0.0116Z^{0.0823}}{0.0073Z^{0.1067}}\right)^{3.781}\right] \quad (26)$$

Figure 10 shows the X_{DRX} as a function of the ε at different deformation conditions. As for each curve, as it increased the ε, the X_{DRX} gradually increases along an S shape. At the same temperature, the X_{DRX} descends with the strain rate. At a low strain rate of 0.01 s^{-1}, the X_{DRX} can reach 100% at each temperature when the ε is up to 0.7. As the strain rate enlarges, low temperature obviously declines the DRX degree, especially for the deformation condition of 573 K and 1 s^{-1} (only ~40% at the 0.7 strain).

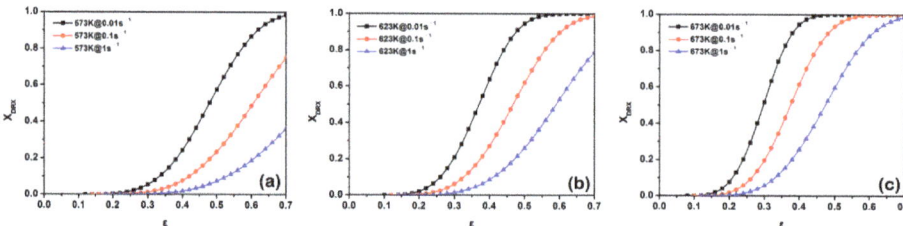

Figure 10. $X_{DRX-\varepsilon}$ curves at (a) 573 K; (b) 623 K; (c) 673 K for 0.01, 0.1 and 1 s^{-1}.

In order to verify the accuracy of the DRX kinetics model, the hot compressive samples at 623 K and 0.01 s^{-1} are selected to observe the DRX area fractions at different strains of 0.3 and 0.5. The result is demonstrated in Figure 11. At 0.3 strain, the DRX and the deformed regions obviously exist (see Figure 11a), and the DRX area fraction is about 36%, as shown in Figure 11c. As the strain increases to 0.5, the DRX region gradually occupies the deformed one (see Figure 11b), and at this moment, the DRX area fraction can reach about 89% shown in Figure 11d. These experiment values are almost accordant with the predicted ones (~30 and ~92%, respectively) using the DRX kinetics model. Therefore, we believe the DRX behavior of the solution-treated BX31 alloy can be forecasted by the DRX kinetics model during hot compression.

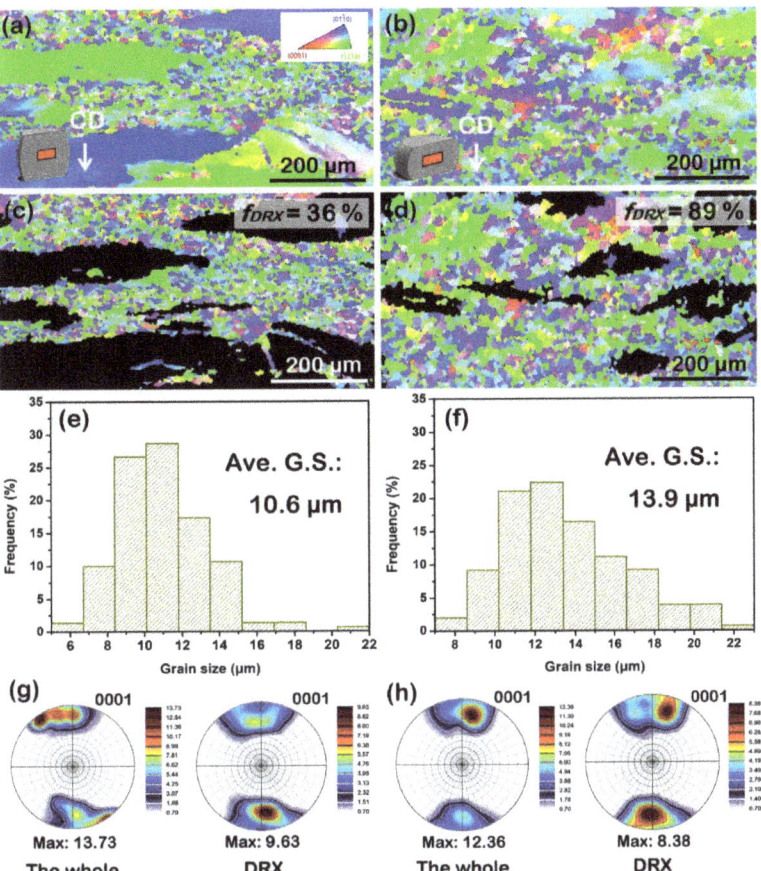

Figure 11. (**a**,**b**) EBSD images of the solution-treated BX31 alloy during the hot compressive strains of 0.3 and 0.5 at 623 K and 0.01 s^{-1}, respectively. CD represents the compressive direction; (**c**,**d**) DRX regions selected from (**a**) and (**b**), respectively; (**e**,**f**) grain size distribution maps of DRX regions from (**c**) and (**d**), respectively; (**g**,**h**) (0001) pole figures containing the whole and DRX areas of the samples with 0.3 and 0.5 strains, respectively.

Furthermore, we also observed that the DRX nucleation preferentially appears at the original grain boundaries. At a high temperature, the CRSSs of non-basal slips may decline, likely activating the non-basal slips within grains that easily induce the obvious orientation gradient. It is also a type of important DRX mechanism (known as continuous DRX) to promote the formation of the DRXed grains, together with the discontinuous DRX mechanisms (i.e., twinning nucleation, PSN, etc.). As the strain increases, the DRXed grains grow, and its average grain size gradually increases from ~10.6 to 13.9 µm, as seen in Figure 11e,f. Generally, the DRX process plays a key role in refining the grains and weakening the basal texture [43,45]. The DRX area shows a lower basal pole intensity than the whole one in the (0001) pole figures for the 0.3 and 0.5 strains (see Figure 11g,h). Interestingly, the subsequent grain growth induces a weaker basal texture. The related literature has stated that the solute segregation (i.e., Ca) into the boundaries of the basal-oriented grains promoted the preferred growth of the non-basal-oriented ones, which might be a crucial factor in weakening the basal texture [33,46].

3.5. Hot Processing Map during Hot Compression

A dynamic material model is valid to describe the hot processing map. In this model, an equation is given [47]:

$$P = G + J = \sigma\dot{\varepsilon} = \int_0^{\dot{\varepsilon}} \sigma d\dot{\varepsilon} + \int_0^{\sigma} \dot{\varepsilon} d\sigma \qquad (27)$$

where P, G and J represent the power dissipated by the workpiece, plastic deformation and microstructural evolution, respectively. The σ as a function of the $\dot{\varepsilon}$ is given [48]:

$$\sigma = K\dot{\varepsilon}^m \qquad (28)$$

where K and m are a material constant and the strain rate sensitivity coefficient, respectively. Together Equation (27) with Equation (28), J is expressed as follows:

$$J = \sigma\dot{\varepsilon} - \int_0^{\dot{\varepsilon}} K\dot{\varepsilon}^m = \frac{m}{m+1}\sigma\dot{\varepsilon} \qquad (29)$$

The efficiency of power dissipation (η) is described as follows [35]:

$$\eta = \frac{J}{J_{max}} = \frac{m\sigma\dot{\varepsilon}/(m+1)}{\sigma\dot{\varepsilon}/2} = \frac{2m}{m+1} \qquad (30)$$

Based on the extreme principle, the instability criterion is proposed as follows [27]:

$$\xi(\dot{\varepsilon}) = \frac{\partial \ln\left(\frac{m}{m+1}\right)}{\partial \ln \dot{\varepsilon}} + m < 0 \qquad (31)$$

Figure 12 shows the hot processing maps at the 0.3, 0.5 and 0.7 strains. Note that the shaded regions represent the flow instability ones. At 0.3 strain, the flow instability region locates at 573~590 K and 0.1~1 s^{-1}, as shown in the top left corner of Figure 12a. When the strain is up to 0.5, two flow instability regions exist. As for the first region, the temperature and strain rate expand to ~630 K and ~0.5 s^{-1} (shown in the top left corner of Figure 12b), and the new-formed second one is between 650~673 K and 0.2~0.6 s^{-1} (seen in the bottom right corner of Figure 12b). As the strain increases to 0.7, the flow instability area at the top left corner almost never changes, while the second one at the bottom right corner gradually enlarges to 630~673 K and 0.2~0.7 s^{-1} in Figure 12c. Based on the stress–strain curves in Figure 3, the solution-treated BX31 alloy lies in a steady-state flow stage after the 0.7 strain, and thus an appropriate hot processing range can be made up by using the result of Figure 12c.

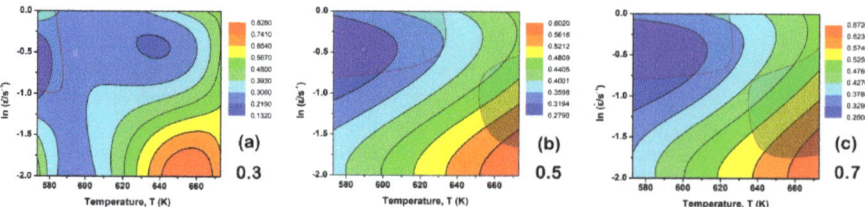

Figure 12. Hot processing maps at different strains: (**a**) 0.3; (**b**) 0.5; (**c**) 0.7.

3.6. Analysis of the as-Extruded BX31 Alloy

According to the hot processing range, we can select a suitable extrusion parameter. It has been widely reported that reducing the extrusion temperature or die-exit speed could effectively obtain fine grains to improve the mechanical properties of Mg alloys [15,21,49]. Therefore, 573 K and 0.1 s^{-1} are selected as the deformation parameters within the range for hot processing to perform the extrusion process. The average strain rate ($\bar{\dot{\varepsilon}}$) is a function of die-exit speed (V_R) at a given extrusion ratio (E_R), billet diameter (D_B) and extrudate diameter (D_E) as follows: [50]:

$$\bar{\dot{\varepsilon}} = \frac{6D_B{}^2 V_R ln E_R}{D_B{}^3 - D_E{}^3} \tag{32}$$

In this work, E_R, D_B and D_E were set as 21:1, 85 and 22 mm, respectively. Thus, the corresponding extrusion parameters contain the extrusion temperature of 573 K and the die-exit speed of 0.5 mm/s.

The solution-treated BX31 alloy is perfectly extruded into a bar at a given deformation condition. Figure 13a shows the macroscopic morphology of the as-extruded BX31 alloy. It can be seen that its surface is very smooth by the observation of high magnification. Obviously, the selected extrusion condition is fitted for the forming of the solution-treated BX31 alloy. Furthermore, the grain structure and micro-texture feature of the as-extruded BX31 alloy are measured, as shown in Figure 13b–d. According to the EBSD result from the longitudinal section, the as-extruded BX31 alloy exhibits a relatively homogeneous and complete DRX characteristic, and its average grain size is ~4.56 μm. In terms of micro-texture, a typical RE texture orientation of [2-1-11]//ED is found on the inverse pole figure, accompanied by the maximum pole intensity of ~2.39. Such a texture was also obtained in other wrought Ca-containing Mg alloys [17,18,51,52]. Ca is considered a special element, such as RE ones, being added into Mg alloys to change the stacking fault energies of non-basal slips, thereby to some extent promoting the activation of non-basal dislocations during extrusion. This may be a crucial factor for a typical RE texture formation in the as-extruded BX31 alloy. The detailed DRX mechanism for this alloy during extrusion can be revealed in future work.

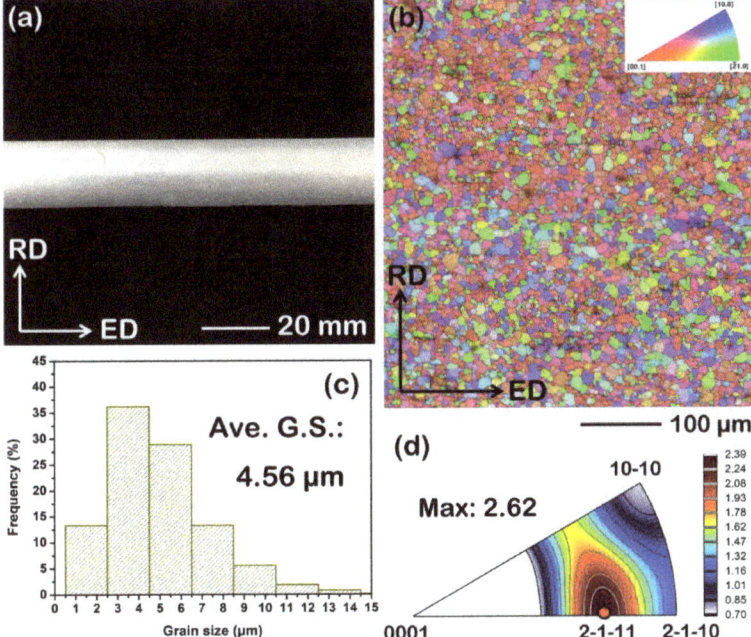

Figure 13. Microstructure analysis of the as-extruded BX31 alloy from longitudinal section (ED-RD plane, RD represents radial direction): (a) macroscopic morphology; (b) EBSD map; (c) grain size distribution map; (d) inverse pole figure.

Additionally, the SEM analysis of the as-extruded BX31 alloy is shown in Figure 14. There exist some granular-shaped phases distributed along ED, and their average sizes are ~2 μm (see Figure 14a,b). Based on the SEM-EDS mapping scanning results shown in Figure 14c–e, these phases mainly also consist of Mg, Bi and Ca elements. Combined with the SEM-EDS point scanning result from phase 1 marked by red arrow in Figure 14b, these phases also can be considered as Mg_2Bi_2Ca ternary phases. It indicates that they belong to the broken remains from the second phase in the solution-treated BX31 alloy after extrusion. These fine second phases accelerate the DRX nucleation, known as the PSN effect.

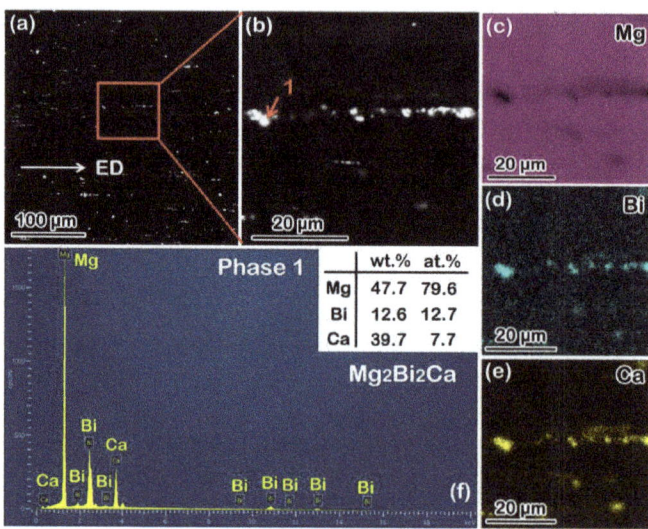

Figure 14. SEM analysis of the as-extruded BX31 alloy from longitudinal section: (**a**) SEM image; (**b**) high magnified view from red frame in (**a**); (**c–e**) SEM-EDS mapping scanning results containing Mg, Bi and Ca elements from (**b**); (**f**) SEM-EDS point scanning result from phase 1 marked by red arrow in (**b**).

Figure 15a shows the RT tensile engineering stress–strain curve and the corresponding tensile data of the as-extruded BX31 alloy. It exhibits a relatively high yield strength of ~192 MPa, excellent ultimate strength of ~230 MPa and the desired elongation-to-failure of ~22%. Figure 15b plots the relationship between the yield strength and the elongation-to-failure for this work and the reported other Mg-Bi-Ca series alloys [17–20]. They are divided into three groups: (1) high strength but low ductility, (2) strength–ductility synergy and (3) high ductility but low strength. Obviously, the reported other Mg-Bi-Ca series alloys have a strength–ductility trade-off dilemma. In comparison, the as-extruded BX31 alloy provides a good balance. Such excellent mechanical performance is mainly attributed to three aspects: (1) Fine grain size. According to the Hall–Petch law [15,21,49], fine grains can offer a considerable number of grain boundaries, effectively hindering the dislocation motions, thereby enhancing the strength of this alloy; (2) Homogeneous complete DRX structure. On the one hand, the high homogeneity of microstructure to some extent reduces the degree of stress concentration, and on the other hand, the low residual dislocation density is beneficial for uniform plastic deformation; (3) The formation of off-basal texture. Generally, weakening basal texture benefits the improvement of the ductility in Mg alloys [17,18,52,53]. In this work, the off-basal orientation of [2-1-11]//ED in the as-extruded BX31 alloy induces a high Schmid factor for basal slip (~0.37 from the calculation of EBSD data) during tension. The activation of more basal slips accommodates the larger strains, which is a key factor in increasing the work-hardening capacity and the ductility of this alloy.

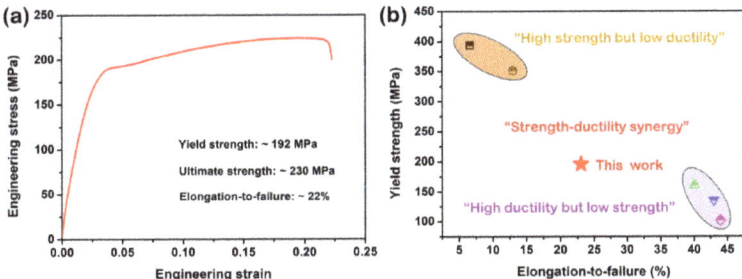

Figure 15. (a) Tensile engineering stress–strain curve of the as-extruded BX31 alloy; (b) relationship between the ultimate strength and the elongation-to-failure for this work (marked by red star) and the reported other Mg-Bi-Ca series alloys (labeled by different symbols). Note that orange and gray circles represent "high strength but low ductility" and "high ductility but low strength" regions, respectively. Reproduced with permission from Refs. [17–20]. Copyright 2020, 2022, Elsevier, MDPI.

4. Conclusions

In this work, the hot deformation behavior of the solution-treated BX31 alloy was systematically studied. The strain-related Arrhenius constitutive model presented a high predictability for the flow curves during hot compression. The flow curves followed the DRX-dominated softening mechanism. The DRX nucleation preferentially appears at the original grain boundaries, and subsequently, non-basal slips might be activated to promote the continuous DRX at a high deformation temperature. For a given deformation condition, the DRX benefited from the texture weakening, and interestingly, with increasing the strain, the DRX growth further reduced the basal texture intensity. This result might be attributed to the preferred growth of non-basal-oriented grains due to the solute segregation. According to the hot processing map, a suitable extrusion parameter could be selected, and the as-extruded BX31 alloy showed a balance of strength and ductility. Such good comprehensive tensile properties were mainly attributed to the following three aspects: (i) fine grain size, (ii) homogeneous complete DRX structure and (iii) the formation of off-basal texture. The newly developed strong and ductile BX31 alloy will be helpful for enriching low-cost, high-performance wrought Mg alloy series for extensive applications in industries.

Author Contributions: Conceptualization, Q.W. and L.W.; methodology, Q.W., L.W. and H.Z.; software, H.Z. and Y.C.; validation, Q.W. and L.W.; formal analysis, Q.W., L.W. and H.Z.; investigation, L.W., H.Z., Y.C. and S.C.; resources, Q.W.; data curation, L.W. and H.Z.; writing—original draft preparation, Q.W.; writing—review and editing, L.W., H.Z., Y.C. and S.C.; visualization, Q.W.; supervision, Q.W.; project administration, Q.W.; funding acquisition, Q.W. All authors have read and agreed to the published version of the manuscript.

Funding: This research was funded by the National Natural Science Foundation of China (No. 52204407), Natural Science Foundation of Jiangsu Province (No. BK20220595), the Postgraduate Research & Practice Innovation Program of Jiangsu Province (No. SJCX22_1720), and the Innovative Science and Technology Platform Project of Cooperation between Yangzhou City and Yangzhou University, China (No. YZ2020266). The APC was funded by the National Natural Science Foundation of China (No. 52204407).

Institutional Review Board Statement: Not applicable.

Informed Consent Statement: Not applicable.

Data Availability Statement: Data available on request.

Conflicts of Interest: The authors declare no conflict of interest.

References

1. Jiang, B.; Dong, Z.; Zhang, A.; Song, J.; Pan, F. Recent advances in micro-alloyed wrought magnesium alloys: Theory and design. *Trans. Nonferrous Met. Soc. China* **2022**, *32*, 1741–1780. [CrossRef]
2. Feng, X.Y.; Jiang, L.J.; Li, Z.N.; Wang, S.M.; Ye, J.H.; Wu, Y.F.; Yuan, B.L. Boosting the hydrogenation activity of dibenzyltoluene catalyzed by Mg-based metal hydrides. *Int. J. Hydrogen Energy* **2022**, *47*, 23994–24003. [CrossRef]
3. Pluengphon, P.; Tsuppayakorn-aek, P.; Tubtimtae, A.; Inceesungvorn, B.; Bovornratanaraks, T. Lightweight alkali-induced lattice-dynamical stability and energy storage mechanism of quaternary Mg-based hydrides under pressure effect. *Int. J. Hydrogen Energy* **2022**, *47*, 30592–30601. [CrossRef]
4. Tong, L.B.; Chu, J.H.; Sun, W.T.; Xu, C.; Zou, D.N.; Wang, K.S.; Kamado, S.; Zheng, M.Y. Achieving an ultra-high strength and moderate ductility in Mg–Gd–Y–Zn–Zr alloy via a decreased-temperature multi-directional forging. *Mater. Charact.* **2021**, *171*, 110804. [CrossRef]
5. Zhang, D.; Yang, Q.; Li, B.; Guan, K.; Wang, N.; Jiang, B.; Sun, C.; Zhang, D.; Li, X.; Cao, Z.; et al. Improvement on both strength and ductility of Mg−Sm−Zn−Zr casting alloy via Yb addition. *J. Alloys Compd.* **2019**, *805*, 811–821. [CrossRef]
6. Rakshith, M.; Seenuvasaperumal, P. Review on the effect of different processing techniques on the microstructure and mechanical behaviour of AZ31 Magnesium alloy. *J. Magnes. Alloy.* **2021**, *9*, 1692–1714. [CrossRef]
7. Zhang, Z.; Zhang, J.; Xie, J.; Liu, S.; He, Y.; Guan, K.; Wu, R. Developing a low-alloyed fine-grained Mg alloy with high strength-ductility based on dislocation evolution and grain boundary segregation. *Scr. Mater.* **2022**, *209*, 114414. [CrossRef]
8. Zhang, D.; Pan, H.; Li, J.; Xie, D.; Zhang, D.; Che, C.; Meng, J.; Qin, G. Fabrication of exceptionally high-strength Mg-4Sm-0.6Zn-0.4Zr alloy via low-temperature extrusion. *Mater. Sci. Eng. A* **2022**, *833*, 142565. [CrossRef]
9. Pan, H.; Kang, R.; Li, J.; Xie, H.; Zeng, Z.; Huang, Q.; Yang, C.; Ren, Y.; Qin, G. Mechanistic investigation of a low-alloy Mg–Ca-based extrusion alloy with high strength–ductility synergy. *Acta Mater.* **2020**, *186*, 278–290. [CrossRef]
10. Xie, D.; Pan, H.; Li, M.; Li, J.; Ren, Y.; Huang, Q.; Yang, C.; Ma, L.; Qin, G. Role of Al addition in modifying microstructure and me-chanical properties of Mg-1.0 wt% Ca based alloys. *Mater Charact.* **2020**, *169*, 110608. [CrossRef]
11. Zeng, Z.R.; Zhu, Y.M.; Nie, J.F.; Xu, S.W.; Davies, C.H.J.; Birbilis, N. Effects of Calcium on Strength and Microstructural Evolution of Extruded Alloys Based on Mg-3Al-1Zn-0.3Mn. *Metall. Mater. Trans. A* **2019**, *50*, 4344–4363. [CrossRef]
12. Zhang, A.; Kang, R.; Wu, L.; Pan, H.; Xie, H.; Huang, Q.; Liu, Y.; Ai, Z.; Ma, L.; Ren, Y.; et al. A new rare-earth-free Mg-Sn-Ca-Mn wrought alloy with ultra-high strength and good ductility. *Mater. Sci. Eng. A* **2019**, *754*, 269–274. [CrossRef]
13. Peng, P.; Tang, A.; Wang, B.; Zhou, S.; She, J.; Zhang, J.; Pan, F. Achieving superior combination of yield strength and ductility in Mg–Mn–Al alloys via ultrafine grain structure. *J. Mater. Res. Technol.* **2021**, *15*, 1252–1265. [CrossRef]
14. Shao, L.; Zhang, C.; Li, C.; Tang, A.; Liu, J.; Yu, Z.; Pan, F. Mechanistic study of Mg-Mn-Al extrusion alloy with superior ductility and high strength. *Mater. Charact.* **2022**, *183*, 111651. [CrossRef]
15. Wang, H.; Zhang, D.T.; Qiu, C.; Zhang, W.W.; Chen, D.L. Achieving superior mechanical properties in a low-alloyed magnesium alloy via low-temperature extrusion. *Mater. Sci. Eng. A* **2022**, *851*, 143611. [CrossRef]
16. Meng, S.; Xiao, H.; Luo, Z.; Zhang, M.; Jiang, R.; Cheng, X.; Yu, H. A new extruded Mg-6Bi-3Al-1Zn alloy with excellent tensile properties. *Metals* **2022**, *12*, 1159. [CrossRef]
17. Meng, S.; Yu, H.; Fan, S.; Kim, Y.; Park, S.; Zhao, W.; You, B.; Shin, K. A high-ductility extruded Mg-Bi-Ca alloy. *Mater. Lett.* **2020**, *261*, 127066. [CrossRef]
18. Meng, S.; Yu, H.; Li, L.; Qin, J.; Woo, S.K.; Go, Y.; Kim, Y.M.; Park, S.H.; Zhao, W.; Yin, F.; et al. Effects of Ca addition on the microstructures and mechanical properties of as-extruded Mg–Bi alloys. *J. Alloys Compd.* **2020**, *834*, 155216. [CrossRef]
19. Meng, S.; Zhang, M.; Xiao, H.; Luo, Z.; Yu, W.; Jiang, R.; Cheng, X.; Wang, L. A superior high-strength dilute Mg-Bi-Ca extrusion alloy with a bimodal microstructure. *Metals* **2022**, *12*, 1162. [CrossRef]
20. Meng, S.; Yu, H.; Han, H.; Feng, J.; Huang, L.; Dong, L.; Nan, X.; Li, Z.; Park, S.H.; Zhao, W. Effect of multi-pass caliber rolling on dilute extruded Mg-Bi-Ca alloy. *Metals* **2020**, *10*, 332. [CrossRef]
21. Wang, Q.; Zhai, H.; Liu, L.; Xia, H.; Jiang, B.; Zhao, J.; Chen, D.; Pan, F. Novel Mg-Bi-Mn wrought alloys: The effects of extrusion temperature and Mn addition on their microstructures and mechanical properties. *J. Magnes. Alloy.* **2022**, *10*, 2588–2606. [CrossRef]
22. Wang, Q.; Zhai, H.; Liu, L.; Jin, Z.; Zhao, L.; He, J.; Jiang, B. Exploiting an as-extruded fine-grained Mg-Bi-Mn alloy with strength-ductility synergy via dilute Zn addition. *J. Alloys Compd.* **2022**, *924*, 166337. [CrossRef]
23. Luo, Y.h.; Cheng, W.l.; Li, H.; Yu, H.; Wang, H.X.; Niu, X.F.; Wang, L.F.; You, Z.Y.; Hou, H. Achieving high strength-ductility synergy in a novel Mg−Bi−Sn−Mn alloy with bimodal microstructure by hot extrusion. *Mater. Sci. Eng. A* **2022**, *834*, 142623. [CrossRef]
24. Hao, J.; Zhang, J.; Xu, C.; Nie, K. Optimum parameters and kinetic analysis for hot working of a solution-treated Mg-Zn-Y-Mn magnesium alloy. *J. Alloys Compd.* **2018**, *754*, 283–296. [CrossRef]
25. Wang, C.; Liu, Y.; Lin, T.; Luo, T.; Zhao, Y.; Yang, Y. Hot compression deformation behavior of Mg-5Zn-3.5Sn-1Mn-0.5Ca-0.5Cu alloy. *Mater. Charact.* **2019**, *157*, 109896. [CrossRef]
26. Sabat, R.K.; Brahme, A.P.; Mishra, R.K.; Inal, K.; Suwas, S. Ductility enhancement in Mg-0.2%Ce alloys. *Acta Mater.* **2018**, *161*, 246–257. [CrossRef]
27. Zhu, G.M.; Wang, L.Y.; Zhou, H.; Wang, J.H.; Shen, Y.; Tu, P.; Zhu, H.; Liu, W.; Jin, P.P.; Zeng, X.Q. Improving ductility of a Mg alloy via non-basal <a> slip induced by Ca addition. *Int. J. Plast.* **2019**, *120*, 164–179. [CrossRef]

28. Zhao, J.; Jiang, B.; Wang, Q.; Yuan, M.; Chai, Y.; Huang, G.; Pan, F. Effects of Li addition on the microstructure and tensile properties of the extruded Mg-1Zn-xLi alloy. *Int. J. Miner. Metall. Mater.* **2022**, *29*, 1380–1387. [CrossRef]
29. Li, Z.J.; Wang, J.G.; Yan, R.F.; Chen, Z.Y.; Ni, T.Y.; Dong, Z.Q.; Lu, T.S. Effect of Ce addition on hot deformation behavior and microstructure evolution of AZ80 magnesium alloy. *J. Mater. Res. Technol.* **2022**, *16*, 1339–1352. [CrossRef]
30. Wang, Q.; Jiang, B.; Tang, A.; He, C.; Zhang, D.; Song, J.; Yang, T.; Huang, G.; Pan, F. Formation of the elliptical texture and its effect on the mechanical properties and stretch formability of dilute Mg-Sn-Y sheet by Zn addition. *Mater. Sci. Eng. A* **2019**, *746*, 259–275. [CrossRef]
31. Remennik, S.; Bartsch, I.; Willbold, E.; Witte, F.; Shechtman, D. New, fast corroding high ductility Mg–Bi–Ca and Mg–Bi–Si alloys, with no clinically observable gas formation in bone implants. *Mater. Sci. Eng. B* **2011**, *176*, 1653–1659. [CrossRef]
32. Liu, Y.H.; Cheng, W.L.; Zhang, Y.; Niu, X.F.; Wang, H.X.; Wang, L.F. Microstructure, tensile properties, and corrosion resistance of extruded Mg-1Bi-1Zn alloy: The influence of minor Ca addition. *J. Alloys Compd.* **2020**, *815*, 152414. [CrossRef]
33. Zhao, D.; Zhao, C.; Chen, X.; Huang, Y.; Hort, N.; Gavras, S.; Pan, F. Compressive deformation of as-extruded LPSO-containing Mg alloys at different temperatures. *J. Mater. Res. Technol.* **2022**, *16*, 944–959. [CrossRef]
34. Ma, Z.; Li, G.; Su, Z.; Wei, G.; Huang, Y.; Hort, N.; Hadadzadeh, A.; Wells, M.A. Hot deformation behavior and microstructural evolution for dual-phase Mg–9Li–3Al alloys. *J. Mater. Res. Technol.* **2022**, *19*, 3536–3545. [CrossRef]
35. Qiao, L.; Zhu, J. Constitutive modeling of hot deformation behavior of AlCrFeNi multi-component alloy. *Vacuum* **2022**, *201*, 111059. [CrossRef]
36. Du, Y.Z.; Liu, D.J.; Ge, Y.F.; Jiang, B.L. Effects of deformation parameters on microstructure and texture of Mg–Zn–Ce alloy. *Trans. Nonferrous Met. Soc. China* **2020**, *30*, 2658–2668. [CrossRef]
37. Wang, T.; Chen, Y.; Ouyang, B.; Zhou, X.; Hu, J.; Le, Q. Artificial neural network modified constitutive descriptions for hot de-formation and kinetic models for dynamic recrystallization of novel AZE311 and AZX311 alloys. *Mater. Sci. Eng. A* **2021**, *816*, 141259. [CrossRef]
38. Park, S.S.; You, B.S.; Yoon, D.J. Effect of the extrusion conditions on the texture and mechanical properties of indirect-extruded Mg-3Al-1Zn alloy. *J. Mater. Process Technol.* **2009**, *209*, 5940–5943. [CrossRef]
39. Du, Y.; Jiang, B.; Ge, Y. Effects of precipitates on microstructure evolution and texture in Mg-Zn alloy during hot deformation. *Vacuum* **2018**, *148*, 27–32. [CrossRef]
40. Savaedi, Z.; Motallebi, R.; Mirzadeh, H. A review of hot deformation behavior and constitutive models to predict flow stress of high-entropy alloys. *J. Alloys Compd.* **2022**, *903*, 163964. [CrossRef]
41. Jeong, H.T.; Park, H.K.; Park, K.; Na, T.W.; Kim, W.J. High-temperature deformation mechanisms and processing maps of equiatomic CoCrFeMnNi high-entropy alloy. *Mater. Sci. Eng. A* **2019**, *756*, 528–537. [CrossRef]
42. Ge, G.; Wang, Z.; Zhang, L.; Lin, J. Hot deformation behavior and artificial neural network modeling of β-γ TiAl alloy containing high content of Nb. *Mater. Today Commun.* **2021**, *27*, 102405. [CrossRef]
43. Pang, H.; Li, Q.; Chen, X.; Chen, P.; Li, X.; Tan, J. Hot deformation behavior and microstructure evolution of Mg-Gd-Y(-Sm)-Zr alloys. *J. Alloys Compd.* **2022**, *920*, 165937. [CrossRef]
44. Li, L.; Wang, Y.; Li, H.; Jiang, W.; Wang, T.; Zhang, C.-C.; Wang, F.; Garmestani, H. Effect of the Zener-Hollomon parameter on the dynamic recrystallization kinetics of Mg–Zn–Zr–Yb magnesium alloy. *Comput. Mater. Sci.* **2019**, *166*, 221–229. [CrossRef]
45. Zhang, Q.; Li, Q.; Chen, X.; Bao, J.; Chen, Z. Effect of Sn addition on the deformation behavior and microstructural evolution of Mg-Gd-Y-Zr alloy during hot compression. *Mater. Sci. Eng. A* **2021**, *826*, 142026. [CrossRef]
46. Kaur, N.; Deng, C.; Ojo, O.A. Effect of solute segregation on diffusion induced grain boundary migration studied by molecular dynamics simulations. *Comput. Mater. Sci.* **2020**, *179*, 109685. [CrossRef]
47. Prasad, Y.V.R.K.; Gegel, H.L.; Doraivelu, S.M.; Malas, J.C.; Morgan, J.T.; Lark, K.A.; Barker, D.R. Modeling of dynamic material behavior in hot deformation: Forging of Ti-6242. *Metall. Trans. A* **1984**, *15*, 1883–1892. [CrossRef]
48. Cheng, W.; Bai, Y.; Ma, S.; Wang, L.; Wang, H.; Yu, H. Hot deformation behavior and workability characteristic of a fine-grained Mg-8Sn-2Zn-2Al alloy with processing map. *J. Mater. Sci. Technol.* **2019**, *35*, 1198–1209. [CrossRef]
49. Du, S.; Yang, K.; Li, M.; Li, J.; Ren, Y.; Huang, Q.; Pan, H.; Qin, G. Achieving high strength above 400 MPa in conventionally ex-truded Mg-Ca-Zn ternary alloys. *Sci. China Technol. Sci.* **2022**, *65*, 519–528. [CrossRef]
50. Lu, X.; Zhao, G.; Zhou, J.; Zhang, C.; Chen, L.; Tang, S. Microstructure and mechanical properties of Mg-3.0Zn-1.0Sn-0.3Mn-0.3Ca alloy extruded at different temperatures. *J. Alloys Compd.* **2018**, *732*, 257–269. [CrossRef]
51. Chai, Y.; Shan, L.; Jiang, B.; Yang, H.; He, C.; Hao, W.; He, J.; Yang, Q.; Yuan, M.; Pan, F. Ameliorating mechanical properties and reducing anisotropy of as-extruded Mg-1.0Sn-0.5Ca alloy via Al addition. *Prog. Nat. Sci.* **2021**, *31*, 722–730. [CrossRef]
52. Zhou, N.; Zhang, Z.; Dong, J.; Jin, L.; Ding, W. High ductility of a Mg–Y–Ca alloy via extrusion. *Mater. Sci. Eng. A* **2013**, *560*, 103–110. [CrossRef]
53. Wang, Q.; Jiang, B.; Chen, D.; Jin, Z.; Zhao, L.; Yang, Q.; Huang, G.; Pan, F. Strategies for enhancing the room-temperature stretch formability of magnesium alloy sheets: A review. *J. Mater. Sci.* **2021**, *56*, 12965–12998. [CrossRef]

Article

Casting Welding from Magnesium Alloy Using Filler Materials That Contain Scandium

Vadym Shalomeev [1,*], Galyna Tabunshchyk [1,*], Viktor Greshta [1], Kinga Korniejenko [2], Martin Duarte Guigou [3] and Sławomir Parzych [2]

[1] National University Zaporizhzhya Polytechnik, 64 Zhukovs' Kogo Street, 69063 Zaporizhzhya, Ukraine; greshtaviktor@gmail.com
[2] Faculty of Material Engineering and Physics, Cracow University of Technology, Jana Pawła II 37, 31-864 Cracow, Poland; kinga.korniejenko@pk.edu.pl (K.K.); slawomir.parzych@pk.edu.pl (S.P.)
[3] Department of Engineering and Technology, Universidad Católica del Uruguay, B de Octubre 2738, Montevideo 11600, Uruguay; martin.duarte@ucu.edu.uy
* Correspondence: shalomeev@zp.edu.ua (V.S.); galina.tabunshchik@gmail.com (G.T.)

Abstract: Based on the results achieved in systematic studies of structure formation and the formation of multicomponent phases, a scandium-containing filler metal from system alloy Mg-Zr-Nd for welding of aircraft casting was developed. The influence of scandium in magnesium filler alloy on its mechanical and special properties, such as long-term strength at elevated temperatures, was studied by the authors. It is established that modification of the magnesium alloy with scandium in an amount between 0.05 and 0.07% allows a fine-grained structure to be obtained, which increases its plasticity up to 70% and heat resistance up to 1.8 times due to the formation of complex intermetallic phases and the microalloying of the solid solution. Welding of the aircraft castings made of magnesium alloy with scandium-containing filler material allows obtaining a weld with a dense homogeneous fusion zone and the surrounding area without any defects. The developed filler material for welding surface defects (cracks, chips, etc.) formed during operation on aircraft engine bodies makes it possible to restore cast body parts and reuse them. The proposed filler material composition with an improved set of properties for the welding of body castings from Mg-Zr-Nd system alloy for aircraft engines makes it possible to increase their reliability and durability in general, extend the service life of aircraft engines, and obtain a significant economic effect.

Keywords: magnesium alloy; filler material; welding; scandium; microstructure modification; mechanical properties

1. Introduction

The development of the aerospace industry results in a constant increase in the complexity of various mechanisms and assemblies, which leads to an increase in their metal intensity [1,2]. One of the solutions to this problem is the use of lightweight materials with a high complex of special mechanical properties [3,4]. Cast magnesium alloys are one of the lightest structural materials, with a density of approximately 1.73 g/cm^3; their low density is approximately 25% that of steel and 60% that of aluminum [4,5]. Furthermore, they have a high specific strength [6]. This makes them a perfect material for the manufacturing of aerospace engineering and the aircraft industry [6,7]. Magnesium alloys must have good casting properties, a high complex of mechanical characteristics, heat resistance, and good weldability [7,8]. Another advantage of magnesium is its recyclability [8–10].

The complexity of aerospace structures and imperfect casting technology require the usage of additive welding technologies [11,12]. Therefore, there is evident progress in the technological trends that combine casting and welding technologies [13,14]. At the same time, when using magnesium alloys, it is necessary to take into account their following characteristics: rather low thermal conductivity, high ignition temperature, increased deformation and stresses during welding due to the high linear expansion coefficient, high gas absorption, the possibility of crystallization crack formation due to fusible eutectics, and large crystallization temperature interval [15,16].

However, magnesium alloys have a lot of advantages, and their applications are also connected to some challenges, including susceptibility to corrosion, low absolute strength, difficulty with deformation, weak chemical stability, flammability, and high price [9,17,18]. Therefore, for engineering and structural applications, magnesium alloys are modified to strengthen their weaker properties without sacrificing their key features [9,19]. The possible methods of modification are alloying, heat treatment, a new preparation method, manufacturing processes, and the application of some additives to magnesium alloys [18,19]. The main objective of these improvements is usually to improve the chemical stability and mechanical properties of magnesium alloys [18,20,21]. The most popular reinforcement that has been studied in recent years is nano-sized particles [9,19] and rare elements [18,20,22]. The additives of rare elements improve the strength and corrosion resistance of magnesium alloys, despite changing the structure of the metal [5,18,20]. However, some investigations have been provided in this area, although this topic requires further research [7,18]. One of the promising additives in this group is scandium [20,23,24].

Alloys of the Mg-Zr-Nd system are promising materials for the production of various molded parts in aerospace engineering [25,26]. They have a fine-grained structure, high physical and mechanical properties, and heat resistance, but have a tendency to crystallize cracks [26,27]. Usually, products made of such magnesium alloys by TIG welding are welded by a non-consumable electrode in an inert gas medium with the use of filler material. The welding filler material is made from the same alloy. This welding technique does not produce a high-quality weld. Defects such as micro-cracks and micro-pores can appear in the weld and seam area [25,28].

The most effective approach in solving the task of increasing the technological strength of welded joints of cast parts made of magnesium alloys is the development of new additive scandium-containing materials with an increased level of mechanical properties and heat resistance, which provide a decrease in the level of residual stresses and the prevention of cracking [29]. The positive effect of scandium is provided on the one hand by the formation of complexly alloyed intermetallic phases, which are additional crystallization centers and refine the grain. On the other hand, the alloy hardening occurs due to the scandium microalloying of the solid solution. Based on global experience, scandium alloys have a high level of mechanical properties and heat resistance; in addition, the welded joints of these alloys are characterized by improved corrosion resistance [23,30]. The usage of scandium alloys is currently constrained by their relatively high cost. However, the requirement to reduce the weight and metal consumption of structures allows the prediction of their widespread use in the near future and ensures the creation of a unified welding technology of complex alloyed magnesium alloys with an improved set of properties [31–33]. The solution to this urgent problem is the central task of research.

The purpose of the research is to develop a filler material for welding surface defects (cracks, chips, etc.) formed during operation on aircraft engine cases. This enables the restoration of cast body parts and their reuse to ensure reliable and durable operation.

2. Materials and Methods

2.1. Materials

The magnesium alloy, with a nominal composition given in Table 1, was smelted in an IPM-500 induction crucible using serial technology.

Table 1. The elemental composition of used magnesium alloy.

Element	Mg	Zr	Nd	Zn
% by mass	96.0%	0.8%	2.6%	0.6%

The refining of the alloy was performed with flux VI-2 in a distribution furnace with batch selection of the melt, which introduced increasing additives of scandium (from 0.05 to 1.0% wt.) Magnesium–scandium ligature (10% Sc, 90% Mg) and standard samples for mechanical tests Ø 12 mm in sand–clay form. The samples were heat treated in Bellevue and PAP-4M furnaces according to the mode: hardening from 415 ± 5 °C, holding time 15 h, cooling in air and aging at 200 ± 5 °C, holding time 8 h, cooling in air.

2.2. Research Methods

Temporary tensile strength and elongation of the samples were determined on a P5 rupture machine P5 at room temperature. The long-term strength at different temperatures was determined on an AIMA 5-2 bursting machine(Russia) on samples Ø 5 mm.

To study the weldability of metal heat-treated plates measuring 200 × 100 × 10 mm of the experimental magnesium alloy, we welded with filler material in the form of cast electrodes Ø 8 × 200 mm from the same alloy containing scandium (between 0.06 and 0.07% wt.). Welding was performed with a non-expendable tungsten electrode in argon medium using a welding transformer TD-500 (Kharkiv, Ukraine), oscillator OSPP-3 (Kiev, Ukraine), ballast rheostat RB-35 (Ukraine), and an argon flow rate between 14 and 18 l/min. For mechanical tests, cylindrical samples Ø 5 mm were made so that the boundary of the transition zone from the weld to the base metal was in the middle of the samples. This limit was detected by etching with a reagent consisting of 1% nitric acid, 20% acetic acid, 19% distilled water, and 60% ethylene glycol.

The quality of the weld was monitored by X-ray. Micro-X-ray spectral analysis of the phases was performed on an electron microscope JSM-6360LA (Tokyo, Japan). The microstructure of the metal was studied under a microscope "Neophot 32". The microhardness of the structural components of the alloy was determined on a microhardness tester company "Buehler" at a load of 0.1 N.

3. Results

3.1. Microstructure Analysis

The microstructure of the initial heat-treated alloy was a δ-solid solution with the presence of a eutectic (δ + γ (MgZr12Nd)) spherical shape (Figure 1a). With increasing scandium content in the alloy, the size of the eutectic increased (Figure 1b,c).

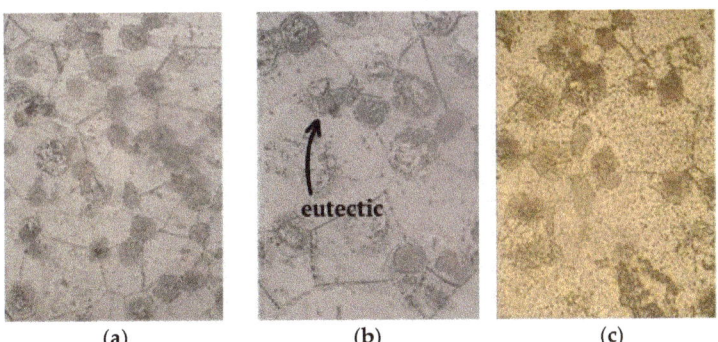

Figure 1. Microstructure of the heat-treated alloy ML10, magnification 500×: (**a**) without Sc; (**b**) 0.05% Sc; (**c**) 1.0% Sc.

Thus, when more than 0.07% Sc was introduced into the melt, the size of the eutectic regions increased approximately three times compared to the initial alloy, while the grain size of the matrix (δ) was practically at the same level (Figure 2). Such an increase in the size of the eutectic phase is due to the presence of an additional amount of fine intermetallic compounds containing scandium.

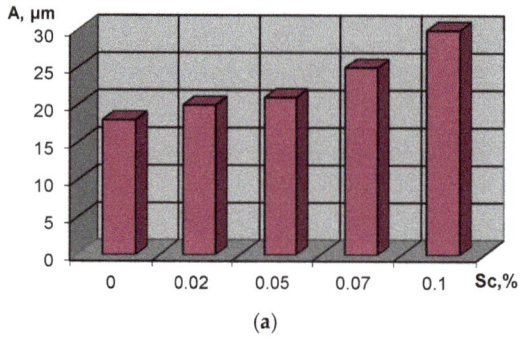
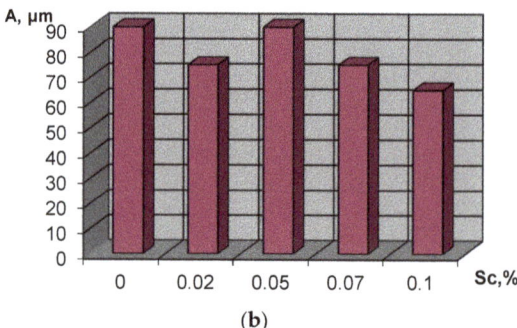

Figure 2. The average size of the structural components: (**a**) of the magnesium alloy with different scandium content: a-(δ + γ)–phase (a eutectic), (**b**) -(δ)–phase (solid solution).

3.2. Microhardness

The microhardness of the structural components of the experimental alloy increased with increasing scandium content, both before and after heat treatment. Moreover, after heat treatment, there was an increase in the microhardness of the matrix and a decrease in the values of the hardness of the eutectic (Table 2), which is associated with the redistribution of scandium between structural components.

Table 2. Microhardness of samples of magnesium alloy with different contents of Sc.

Amount of Sc (% by wt.)	Average Microhardness HV (MPa)			
	To Heat Treatment		After Heat Treatment	
	Matrix	Eutectic	Matrix	Eutectic
-	662.5	1916.6	1065.7	1320.4
0.02	770.0	2089.3	1078.8	1480.6
0.05	781.0	2094.9	1098.8	1504.7
0.07	827.1	2125.7	1114.4	1735.6
0.10	835.0	2158.7	1154.5	1891.6
0.30	846.5	2211.8	1187.4	1930.6
0.50	871.5	2285.9	1235.5	1985.7
0.70	904.3	2348.3	1288.4	2130.6
1.00	942.8	2450.7	1320.5	2211.6

3.3. Micro-X-ray Spectral Analysis

Micro-X-ray spectral analysis of the eutectic showed that it has an elemental composition, as given in Table 3.

Table 3. The obtained results of Micro-X-ray spectral analysis.

Element	Mg	Zr	Nd	Si	Sc
% by mass	93.52	1.83	4.0	0.08	0.57

The scandium content in the eutectic was between 1.5 and 2.0 times higher than in the δ-solid solution. Thus, doping of the magnesium alloy with scandium leads, first of all, to the saturation of the eutectic phase. This is confirmed by the increase in the microhardness of the eutectic with the increase in scandium additives in the alloy.

When the scandium content in the alloy is not more than 0.05%, there was an increase in both mechanical and heat-resistant properties due to the microalloying of structural components. A further increase in the concentration of scandium in the metal led to an increase in the intermetallic phase, redistribution of excess intermetallics at the grain boundary, and reduced the physical and mechanical characteristics of the material (Table 4).

Table 4. Average mechanical properties and long-term strength of magnesium alloy with different contents of Sc.

Amount of Sc (% by wt.)	Mechanical Properties		Long-Term Strength at σ = 80 MPa, Hours		
	σ_y (MPa)	δ (%)	T [1] = 150/250 °C	T = 270 °C	T = 300 °C
0	235.0	3.6	1252.0/2.,2	47.5	9.0
0.02	253.0	4.6	1252.0/56.0	53.1	11.1
0.05	245.0	6.3	1252.0/48.7	71.5	16.0
0.07	240.0	4.0	1252.0/64.0	61.6	12.4
0.10	232.0	3.5	1252.0/48.0	36.5	13.4
0.50	235.0	4.0	1251.0/34.1	24.0	6.7
1.00	169.0	3.3	1252.0/8.0	-	-

[1] Tests of samples for long-term strength were performed in stages: at 150 °C (numerator), then at 250 °C (denominator).

3.4. The Study the Weldability

The positive effect of scandium on the long-term strength of the alloy at elevated temperatures was noted. The increase in the content of scandium in the alloy, leading to microalloying of its structural components, helped to increase its heat resistance. Increasing the temperature during the long-term strength test reduced the time to failure. Fine intermetallic particles were released unevenly, forming areas of a striped structure with high microhardness, leading to the destruction of the metal.

In the structure of samples containing more than 0.07% scandium, coarse boundary inclusions were observed, leading to the rapid destruction of samples during testing. Thus, the content between 0.05 and 0.06% of the scandium mass in the experimental magnesium alloy is optimal for the manufacture of filler material with improved properties.

Qualitative indicators of Mg-Zr-Nd alloy castings welded with filler material with scandium content between 0.05 and 0.06%wt., including the weld and close-seam region, were studied. The structure of the seam area is characteristic of the heat-treated alloy. In the structure of the weld, δ-solid solution and γ-phase located along the grain boundaries were observed in the form of light gray emissions.

The dimensions of the structural components of the weld in comparison with the base metal were significantly smaller (Table 5), and the microhardness of the weld was slightly higher.

Table 5. The average size of structural components, microhardness, and mechanical properties of welded samples of magnesium alloy.

Amount of Sc (% by wt.)	Average Size of Structural Components		Mechanical Properties	
	Matrix (µm)	Eutectic (µm)	σ_y (MPa)	δ (%)
0	70/34	40/30	235/239	3.6/3.2
0.05–0.06	60/25	45/35	245/253	5.6/6.0

Note: the numerator is the base metal, and the denominator is the weld.

Mechanical tests of samples with a weld showed that the destruction of the metal took place in the near-seam zone because the weld metal had a fine-grained structure due to accelerated crystallization, which positively contributed to the improvement of the quality of the test metal.

An industrial test of the additive alloy with scandium at welding was carried out during the welding of magnesium alloy body casting elements. There was conducting welding of elements of fuel pump housings, low-pressure compressor housings, and gearbox housings, which are integral parts of aircraft engines. After, these areas were cleaned until the complete removal of defects and thoroughly degreased. The prepared products were placed in a thermal oven, heated to 200 °C, kept at this temperature for 6 h, removed from the oven, and welded. Welding was performed by an argon-arc method using a non-consumable tungsten electrode and a welding transformer TD-500, oscillator OSPP-3, and ballast rheostat RB-35. Rods Ø 8 × 200 mm made of the studied magnesium alloy with scandium were used as the filler material. The welded parts were again loaded into a thermal oven and cooled together with the oven to a temperature not exceeding 120 °C. The welding areas were monitored and cleaned to obtain the required geometric dimensions.

Welded body casting elements were cut at the welding sites to determine the quality of the metal. Metallographic and X-ray inspection showed that the weld and the base metal (Figure 3) had a dense homogeneous structure without defects. The weld microstructure consisted of a δ-solid solution and intermetallic phase containing Sc, Zr, and Nd.

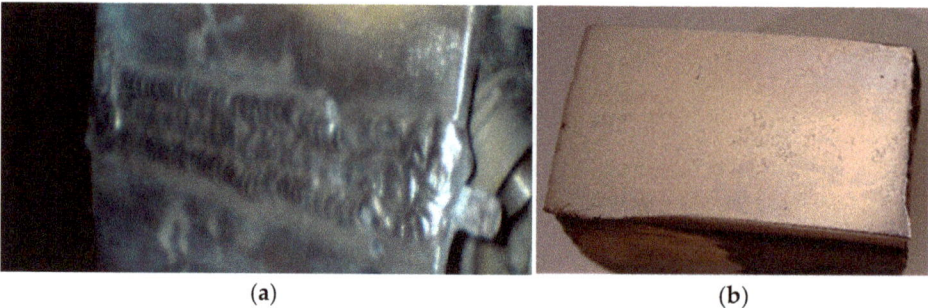

(a) (b)

Figure 3. Magnesium alloy body casting element after welding: (a) magnification 0.1×; (b) macrograph.

In this case, the size of the structural components in the weld was much smaller than in the main metal (Figure 4). The weld metal did not contain any defects (micropores, microcracks, etc.) and corresponded to the base metal in terms of mechanical properties.

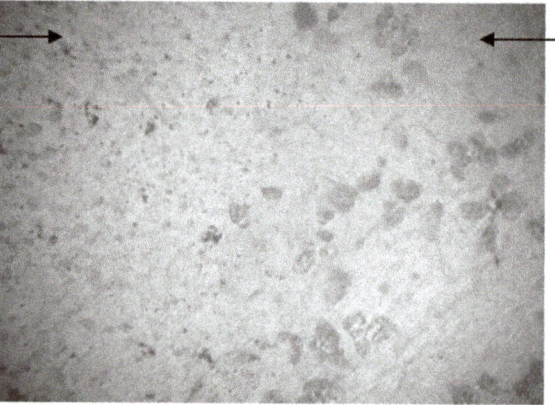

Figure 4. Microstructure of magnesium alloy near the weld zone, magnification 200×.

The developed technology of welding castings from a magnesium alloy enables the restoration of difficult case details of aircraft engines in which qualitative indicators satisfy the requirements of regulatory and technical documentation.

4. Discussion

This investigation showed the possibility of using magnesium alloys to produce elements for the aerospace industry. The additive of scandium optimized the microstructure of the alloy and increased its properties. Other studies in the literature confirm the presented results, especially the strong influence of scandium on the microstructure of magnesium alloys [23]. Li et al. [23] showed that a scandium additive of up to 1% wt. refined the grain sizes of the material. They defined the main mechanism of this phenomenon as the heterogeneous nucleation mechanism [23].

Additionally, the mechanical properties of the magnesium alloy with scandium were changed, including the improvement in yield [23,34]. The investigation provided by Li et al. [34] confirmed the improvement in the Vickers hardness magnesium alloys with the addition of scandium compared to pure magnesium alloys. One of the implications of this research is the conclusion that the addition of scandium could enhance the wear resistance of magnesium alloys [34].

Other research shows that scandium can improve magnesium alloy's properties, including corrosion resistance [35,36]. This fact can support the application of this material not only in the aerospace industry but also in other areas, such as implant applications [37,38].

It is worth noting that a similar mechanism was also observed in aluminum–magnesium alloys as well as other alloys that contain magnesium [39,40].

Another challenge is to study scandium as an additive in the context of material weldability, which is not a trivial issue [8,41]. In the literature, there are only a few studies in this area. Most of them are related to the development of additive manufacturing technologies [5,42,43]. The obtained results show that the addition of scandium could have a positive influence on the development of welding technology as well as additive manufacturing for magnesium alloys and help overcome challenges connected with processing magnesium alloys [5,44].

Magnesium-based alloys that contain zirconium and neodymium, which form heat-resistant intermetallic phases and provide the required service characteristics of the alloy at elevated temperatures, are widely used for the production of heat-resistant magnesium castings in aircraft engines. The range of cast parts made from this alloy is quite diverse and involves a combination of different technologies for their manufacturing (casting, welding, heat treatment, etc.). Welding of the products made of the Mg-Zr-Nd alloy system is carried out with the use of a filler material and an alloy base. In this case, parts of the complex configuration are welded poorly; often, microcracks are formed in the welding spots, which require re-welding. After several unsuccessful welding operations, this product is rejected and sent for remelting. One way to solve this problem is to develop a new composition of a filler material with higher mechanical and special properties compared to the base metal. Therefore, it is important to study the effect of scandium on the structure and properties of the alloy of the Mg-Zr-Nd system, which already has heat-resistant phases (MgZr) 12Nd in its composition, which will improve the physical and mechanical characteristics of the metal in the place of welding.

The research was carried out in two stages:

1. The development of a magnesium-based additive alloy with optimal scandium content to ensure improved mechanical properties and heat resistance.
2. The investigation of the structure and properties of the base metal and the weld on samples welded with the developed scandium-containing filler material.

The structure and properties of a magnesium alloy containing neodymium and zirconium and additionally modified with scandium in an amount of up to 1.0% were studied. Metallographic analysis of the investigated metal showed that additives in the alloy contributed to an increase in the size of the spherical areas of the precipitation of eutectoid. Thus, when more than 0.07% Sc was introduced into the melt, the size of the eutectoid areas increased approximately four times compared to the standard alloy, while the size of the δ-phase was approximately at the same level.

Thermal treatment contributes to the homogeneity of the alloy due to the redistribution of elements between the dendrite axes and interaxial spaces, as well as additional alloying of the matrix due to the diffusion of elements from the boundary precipitations of the (MgZr)12Nd phase.

Micro X-ray diffraction analysis showed that the spherical regions were enriched, mainly with zirconium, neodymium, and scandium. In modified alloys, the scandium content in the spherical regions of δ + (MgZr)12Nd eutectoid excretions was ~1.5–2.0 times higher than in the δ-solid solution.

When the content of scandium in the alloy was lower than 0.3%, grain refinement was observed. A further increase in the filler modifier (up to 1.0% Sc) led to an increase in the size of the micrograins up to 160 μm (in the range of 0.02–0.3% Sc, the size of the micrograins is ~75 μm).

Studies on the heat resistance showed that the samples heated to temperatures of 150–250 °C showed the decomposition of eutectoid. Analysis of the microstructures showed that in the process of exposure to temperature and prolonged exposure along with the decomposition of the eutectoid, its dissolution in the matrix with the subsequent release of intermetallic phases of the (MgZr)12Nd type was the form of finely dispersed particles. At the same time, the fine intermetallic particles were isolated irregularly, forming areas of a banded structure characterized by an increase in microhardness values.

It was established that a more complete disintegration of the eutectoid phase was promoted by the holding time at a given temperature, as well as by the stress value. At temperature 270 °C, coarsening of the structure was observed due to intensive allocation of intermetallides, in particular, on borders of grains, which explains the sharp decrease in the heat resistance of the material. The coarsest boundary separation was detected in the structure of samples containing more than 0.07% Sc, leading to the rapid destruction of samples during the long-term strength test.

The δ-solid solution microhardness of the standard alloy (before heat treatment) was more than three times lower than the microhardness of the isolations in the spherical eutectoidal regions. After heat treatment, an increase in matrix microhardness and a decrease in eutectoid hardness values were observed, indicating an increase in the homogeneity of the heat-treated alloy. At the same time, an increase in scandium concentration in the alloy resulted in an increase in microhardness values of structural components both before and after heat treatment.

When increasing the holding time at temperatures of 150–250 °C, the microhardness of the alloys under study decreased due to a more complete decomposition of the eutectoid type δ + (MgZr)12Nd.

The addition of scandium to the alloy up to 0.07% contributed to an increase in both mechanical and heat resistance properties. Further, there was a tendency to decrease the physical and mechanical properties of the material.

Increasing the long-term strength test temperature to 270 °C reduced the time to failure by ~6 times. Samples with an additive of 1.0% Sc were destroyed when they were loaded already at 250 °C because of the formation of microloops and film impurities.

Thus, the scandium content in the metal in the range of 0.05–0.07% contributes to obtaining a finely dispersed, homogeneous structure and provides the best indices of mechanical properties and heat resistance of the alloy.

The developed composition of filler material was tested for the welding of magnesium castings. The structure and properties of the base metal and the weld were studied on the products welded with the developed scandium-containing filler material.

Some of the aircraft engine casings were welded with the use of the experimental filler material. After, they were cut for metallographic control and the production of samples for mechanical tests. Metallographic control of the welded hull elements showed that the weld metal and the base of the part had a dense homogeneous structure without defects. The microstructure of the base metal was characteristic of the heat-treated state; the weld consisted of a δ-solid solution and an intermetallic phase. At the same time, the size of the structural components in the weld was much smaller than in the base metal. To control the mechanical properties, samples were cut from the welded body parts. The samples were taken so that the boundary of the transition zone from the weld to the base metal was in the middle part of the samples. To determine this limit, samples were etched with a reagent consisting of 1% nitric acid, 20% acetic acid, 19% distilled water, and 60% ethylene glycol. X-ray control established good quality of welding. By the level of mechanical properties (σv = 230–235 MPa, δ = 3.5–4.0%), the examined metal exceeded the indices of the base metal, and in that specimen, destruction during tests was not along the weld but along the base metal.

Thus, the usage of scandium-containing filler material for welding complex hulls of aircraft engines made of a magnesium alloy of the Mg-Zr-Nd system allows for a high-quality weld and ensures the reliable operation of critical units. It is worth stressing that similar research is also being developed for others alloys, such as aluminum [45,46], and some obtained results could be an inspiration for others materials.

5. Conclusions

The application of scandium-containing filler material for welding products made of Mg-Zr-Nd alloy system helps improve significant mechanical properties, heat resistance, and the reliability of aircraft structures in general. Moreover, the developed technology for welding cast products made of magnesium alloy with scandium-containing filler enhances their operational characteristics and extends the service life of the aircraft mechanisms and units and results in a significant economic effect. The provided research allows us to formulate the following conclusions:

- The modification of the magnesium alloy with scandium in an amount of scandium between 0.05 and 0.06 of the mass allows us to obtain a fine-grained structure, an increase in the level of mechanical properties, and long-term strength at 250 °C due to the formation of complex intermetallic phases and a microalloying δ-solid solution.
- Welding of elements of cast parts from the alloy system, using Mg-Zr-Nd filler material containing scandium, enabled a dense homogeneous zone of alloying to be obtained with high mechanical properties.
- Industrial testing of the technological process of welding casting elements with scandium filler material allowed us to recommend it in the production process to obtain complex structures that meet operational regulatory requirements.

Author Contributions: Conceptualization, V.S. and G.T.; methodology, V.S., G.T. and V.G.; validation, V.S., G.T. and V.G.; formal analysis, K.K. and S.P.; investigation, V.S., G.T. and V.G.; resources, V.S.; writing—original draft preparation, V.S., G.T. and K.K.; writing—review and editing, V.G., S.P. and M.D.G.; supervision, V.S.; funding acquisition, M.D.G. and K.K. All authors have read and agreed to the published version of the manuscript.

Funding: This work was financed by the Polish National Agency for Academic Exchange under the International Academic Partnership Programme within the framework of the grant: E-mobility and sustainable materials and technologies EMMAT (PPI/APM/2018/1/00027).

Institutional Review Board Statement: Not applicable.

Informed Consent Statement: Not applicable.

Data Availability Statement: Not applicable.

Conflicts of Interest: The authors declare no conflict of interest. The funders had no role in the design of the study; in the collection, analyses, or interpretation of data; in the writing of the manuscript, or in the decision to publish the results.

References

1. Barroqueiro, B.; Andrade-Campos, A.; Valente, R.A.F.; Neto, V. Metal Additive Manufacturing Cycle in Aerospace Industry: A Comprehensive Review. *J. Manuf. Mater. Process.* **2019**, *3*, 52. [CrossRef]
2. Toozandehjani, M.; Kamarudin, N.; Dashtizadeh, Z.; Lim, E.Y.; Gomes, A.; Gomes, C. Conventional and Advanced Composites in Aerospace Industry: Technologies Revisited. *Am. J. Aerosp. Eng.* **2018**, *5*, 9–15. [CrossRef]
3. Bahl, B.; Singh, T.; Kumar, V.; Sehgal, S.; Kumar Bagha, A. A systematic review on recent progress in advanced joining techniques of the lightweight materials. *AIMS Mater. Sci.* **2021**, *8*, 62–81. [CrossRef]
4. Zhu, L.; Li, N.; Childs, P.R.N. Light-weighting in aerospace component and system design. *Propuls. Power Res.* **2018**, *7*, 103–119. [CrossRef]
5. Kurzynowski, T.; Pawlak, A.; Smolina, I. The potential of SLM technology for processing magnesium alloys in aerospace industry. *Archiv. Civ. Mech. Eng.* **2020**, *20*, 23. [CrossRef]
6. Wendt, A.; Weiss, K.; Ben-Dov, A.; Bamberger, M.; Bronfin, B. Magnesium Castings in Aeronautics Applications—Special Requirements. In *Essential Readings in Magnesium Technology*; Mathaudhu, S.N., Luo, A.A., Neelameggham, N.R., Nyberg, E.A., Sillekens, W.H., Eds.; Springer: Cham, Switzerland, 2016; pp. 65–69. [CrossRef]
7. Volkova, E.F. Modern wrought alloys and composite materials based on magnesium. *Met. Sci. Therm. Process. Met.* **2006**, *11*, 5–9. (In Russian)
8. Gloria, A.; Montanari, R.; Richetta, M.; Varone, A. Alloys for Aeronautic Applications: State of the Art and Perspectives. *Metals* **2019**, *9*, 662. [CrossRef]
9. Tan, J.; Ramakrishna, S. Applications of Magnesium and Its Alloys: A Review. *Appl. Sci.* **2021**, *11*, 6861. [CrossRef]
10. Golroudbary, S.R.; Makarava, I.; Repo, E.; Kraslawski, A.; Luukka, P. Magnesium Life Cycle in Automotive Industry. *Procedia CIRP* **2022**, *105*, 589–594. [CrossRef]
11. Sagar, P.; Handa, A. A comprehensive review of recent progress in fabrication of magnesium base composites by friction stir processing technique—A review. *AIMS Mater. Sci.* **2020**, *7*, 684–704. [CrossRef]
12. Blakey-Milner, B.; Gradl, P.; Snedden, G.; Brooks, M.; Pitot, J.; Lopez, E.; Leary, M.; Berto, F.; du Plessis, A. Metal additive manufacturing in aerospace: A review. *Mater. Des.* **2021**, *209*, 110008. [CrossRef]
13. Sereda, B.; Sheyko, S.; Belokon, Y.; Sereda, D. The influence of modification on structure and properties of rapid steel. In Proceedings of the AIST Steel Properties and Applications Conference Proceedings-Combined with MS and T'11, Materials Science and Technology 2011, Columbia, OH, USA, 16 October 2011.
14. Sheyko, S.; Sukhomlin, G.; Mishchenko, V.; Shalomeev, V.; Tretiak, V. Formation of the Grain Boundary Structure of Low-Alloyed Steels in the Process of Plastic Deformation. In Proceedings of the Materials Science and Technology 2018, MS and T 2018, Columbia, OH, USA, 14 October 2018. [CrossRef]
15. Bondarev, A.A.; Nesterenkov, V.M. Technological features of electron beam welding of deformable magnesium alloys in vacuum. *Autom. Weld.* **2014**, *3*, 18–22. (In Russian)
16. Malik, A.; Nazeer, F.; Wang, Y. A Prospective Way to Achieve Ballistic Impact Resistance of Lightweight Magnesium Alloys. *Metals* **2022**, *12*, 241. [CrossRef]
17. Czerwinski, F. Magnesium and Its Alloys. In *Magnesium Injection Molding*; Springer: New York, NY, USA, 2008; pp. 1–79.
18. Luo, Q.; Guo, Y.; Liu, B.; Feng, Y.; Zhang, J.; Li, Q.; Chou, K. Thermodynamics and kinetics of phase transformation in rare earth–magnesium alloys: A critical review. *J. Mater. Sci. Technol.* **2020**, *44*, 171–190. [CrossRef]
19. Gupta, M.; Wong, W.L.E. Magnesium-Based Nanocomposites: Lightweight Materials of the Future. *Mater. Charact.* **2015**, *105*, 30–46. [CrossRef]
20. Sivashanmugam, N.; Harikrishna, K.L. Influence of Rare Earth Elements in Magnesium Alloy—A Mini Review. *Mater. Sci. Forum* **2020**, *979*, 162–166. [CrossRef]
21. Yan, K.; Sun, J.; Liu, H.; Cheng, H.; Bai, J.; Huang, X. Exceptional mechanical properties of an Mg97Y2Zn1 alloy wire strengthened by dispersive LPSO particle clusters. *Mater. Lett.* **2019**, *242*, 87–90. [CrossRef]
22. Zhang, J.; Liu, S.; Wu, R.; Hou, L.; Zhang, M. Recent developments in high-strength Mg-RE-based alloys: Focusing on Mg-Gd and Mg-Y systems. *J. Magnes. Alloys* **2018**, *6*, 277–291. [CrossRef]
23. Li, T.; He, Y.; Zhou, J.; Tang, S.; Yang, Y.; Wang, X. Microstructure and mechanical property of biodegradable Mg–1.5Zn–0.6Zr alloy with varying contents of scandium. *Mater. Lett.* **2018**, *229*, 60–63. [CrossRef]
24. Dovzhenko, N.N.; Demchenko, A.I.; Bezrukikh, A.A.; Dovzhenko, I.N.; Baranov, V.N.; Orelkina, T.A.; Dementeva, I.S.; Voroshilov, D.S.; Gaevskiy, V.N.; Lopatina, E.S. Mechanical properties and microstructure of multi-pass butt weld of plates made of Al-Mg-Zr alloy sparingly doped with scandium. *Int. J. Adv. Manuf. Technol.* **2021**, *113*, 785–805. [CrossRef]
25. Shalomeev, V.; Tsivirco, E.; Vnukov, Y.; Osadchaya, Y.; Makovskyi, S. Development of new casting magnesiumbased alloys with increased mechanical properties. *East.-Eur. J. Enterp. Technol.* **2016**, *4*, 4–10.
26. Li, N.; Liu, J.R.; Wang, S.Q.; Sheng, S.J.; Huang, W.D.; Pang, Y.T. Application of rare earth in magnesium and magnesium alloys. *Foundry Technol.* **2006**, *27*, 1133–1136.
27. Liu, K.; Kou, S. Susceptibility of magnesium alloys to solidification cracking. *Sci. Technol. Weld. Join.* **2020**, *25*, 251–257. [CrossRef]
28. Zhou, W.; Aprilia, A.; Mark, C.K. Mechanisms of Cracking in Laser Welding of Magnesium Alloy AZ91D. *Metals* **2021**, *11*, 1127. [CrossRef]

29. Zhemchuzhnikova, D.; Kaibyshev, R. Effect of Grain Size on Cryogenic Mechanical Properties of an Al-Mg-Sc Alloy. *Adv. Mater. Res.* 2014, *922*, 862–867. [CrossRef]
30. Zhemchuzhnikova, D.; Kaibyshev, R. Effect of Rolling on Mechanical Properties and Fatigue Behavior of an Al-Mg-Sc-Zr Alloy. *Mater. Sci. Forum* 2014, *794–796*, 331–336. [CrossRef]
31. Belikov, S.; Shalomeev, V.; Tsivirko, E.; Aikin, N.; Sheyko, S. Microalloyed magnesium alloys with high complex of properties. In Proceedings of the Materials Science and Technology Conference and Exhibition 2017, MS and T 2017, Pittsburgh, PA, USA, 8 October 2017. [CrossRef]
32. Velikiy, V.I.; Yares'ko, K.I.; Shalomeev, V.A.; Tsivirko, E.I.; Vnukov, Y.N. Prospective magnesium alloys with elevated level of properties for the aircraft engine industry. *Met. Sci. Heat Treat.* 2014, *55*, 492–498. [CrossRef]
33. Filatov, Y.A. Various approaches to the implementation of the strengthening effect from the addition of scandium in wrought alloys based on the Al-Mg-Sc system. *VILS Technol. Light Alloy.* 2009, *3*, 42–45. (In Russian)
34. Li, T.; Wang, X.T.; Tang, S.Q.; Yang, Y.S.; Wu, J.H.; Zhou, J.X. Improved wear resistance of biodegradable Mg–1.5Zn–0.6Zr alloy by Sc addition. *Rare Met.* 2021, *40*, 2206–2212. [CrossRef]
35. Zhao, P.; Xie, T.; Ying, T.; Zhu, H.; Zeng, X. Role of Alloyed Sc on the Corrosion Behavior of Mg. *Metall. Mater. Trans. A Phys. Metall. Mater. Sci.* 2022, *53*, 741–746. [CrossRef]
36. Li, T.; He, Y.; Wu, J.; Zhou, J.; Tang, S.; Yang, Y.; Wang, X. Effects of scandium addition on the in vitro degradation behavior of biodegradable Mg–1.5Zn–0.6Zr alloy. *J. Mater. Sci.* 2018, *53*, 14075–14086. [CrossRef]
37. He, R.; Liu, R.; Chen, Q.; Zhang, H.; Wang, J.; Guo, S. In vitro degradation behavior and cytocompatibility of Mg-6Zn-Mn alloy. *Mater. Lett.* 2018, *228*, 77–80. [CrossRef]
38. Li, T.; He, Y.; Zhou, J.; Tang, S.; Yang, Y.; Wang, X. Effects of scandium addition on biocompatibility of biodegradable Mg–1.5Zn–0.6Zr alloy. *Mater. Lett.* 2018, *215*, 200–202. [CrossRef]
39. Yuryev, P.O.; Baranov, V.N.; Orelkina, T.A.; Bezrukikh, A.I.; Voroshilov, V.S.; Murashkin, M.Y.; Partyko, E.G.; Konstantinov, I.L.; Yanov, V.V.; Stepanenko, N.A. Investigation the structure in cast and deformed states of aluminum alloy, economically alloyed with scandium and zirconium. *Int. J. Adv. Manuf. Technol.* 2021, *115*, 263–274. [CrossRef]
40. Zhang, M.; Yan, H.; Wei, Q. Effects of Scandium Addition on the Structural Stability and Ideal Strengths of Magnesium-Lithium Alloys. *Z. Für Nat. A* 2018, *73*, 947–956. [CrossRef]
41. Łach, M.; Korniejenko, K.; Balamurugan, P.; Uthayakumar, M.; Mikuła, J. The Influence of Tuff Particles on the Properties of the Sintered Copper Matrix Composite for Application in Resistance Welding Electrodes. *Appl. Sci.* 2022, *12*, 4477. [CrossRef]
42. Pouranvari, P. Critical review on fusion welding of magnesium alloys: Metallurgical challenges and opportunities. *Sci. Technol. Weld. Join.* 2021, *26*, 559–580. [CrossRef]
43. Rubino, F.; Parmar, H.; Esperto, V.; Carlone, P. Ultrasonic welding of magnesium alloys: A review. *Mater. Manuf. Process.* 2020, *35*, 1051–1068. [CrossRef]
44. Han, D.; Zhang, J.; Huang, J.; Lian, Y.; He, G. A review on ignition mechanisms and characteristics of magnesium alloys. *J. Magnes. Alloys* 2020, *8*, 329–344. [CrossRef]
45. Saini, N.; Pandey, C.; Dwivedi, K.D. Ductilizing of cast hypereutectic Al–17%Si alloy by friction stir processing. *Proc. Inst. Mech. Eng. Part E J. Process Mech. Eng.* 2017, *232*, 696–701. [CrossRef]
46. Saini, N.; Pandey, C.; Thapliyal, S.; Dwivedi, K.D. Mechanical Properties and Wear Behavior of Zn and MoS2 Reinforced Surface Composite Al- Si Alloys Using Friction Stir Processing. *Silicon* 2018, *10*, 1979–1990. [CrossRef]

Article

Effect of the Ca$_2$Mg$_6$Zn$_3$ Phase on the Corrosion Behavior of Biodegradable Mg-4.0Zn-0.2Mn-xCa Alloys in Hank's Solution

Junjian Fu, Wenbo Du *, Ke Liu *, Xian Du, Chenchen Zhao, Hongxing Liang, Adil Mansoor, Shubo Li and Zhaohui Wang

Faculty of Materials and Manufacturing, Beijing University of Technology, Beijing 100124, China; fujunjian@emails.bjut.edu.cn (J.F.); duxian@bjut.edu.cn (X.D.); zhaochenchen@bjut.edu.cn (C.Z.); hongxingliang314@gmail.com (H.L.); adilmansoor_786@outlook.com (A.M.); lishubo@bjut.edu.cn (S.L.); wangzhaohui@bjut.edu.cn (Z.W.)
* Correspondence: duwb@bjut.edu.cn (W.D.); lk@bjut.edu.cn (K.L.); Tel.: +86-10-67392917 (W.D.); +86-10-67392423 (K.L.)

Abstract: The effect of the Ca$_2$Mg$_6$Zn$_3$ phase on the corrosion behavior of biodegradable Mg-4.0Zn-0.2Mn-xCa (ZM-xCa, x = 0.1, 0.3, 0.5 and 1.0 wt.%) alloys in Hank's solution was investigated with respect to phase spacing, morphology, distribution and volume fraction. With the increase in Ca addition, the volume fraction of the Ca$_2$Mg$_6$Zn$_3$ phase increased from 2.5% to 7.6%, while its spacing declined monotonically from 43 μm to 30 μm. The Volta potentials of secondary phases relative to the Mg matrix were measured by using scanning kelvin probe force microscopy (SKPFM). The results show that the Volta potential of the intragranular spherical Ca$_2$Mg$_6$Zn$_3$ phase (+109 mV) was higher than that of the dendritic Ca$_2$Mg$_6$Zn$_3$ phase (+80 mV). It is suggested that the Ca$_2$Mg$_6$Zn$_3$ acted as a cathode to accelerate the corrosion process due to the micro-galvanic effect. The corrosion preferred to occur around the spherical Ca$_2$Mg$_6$Zn$_3$ phase at the early stage and developed into the intragranular region. The corrosion rate increased slightly with increasing Ca content from 0.1 wt.% to 0.5 wt.% because of the enhanced micro-galvanic corrosion effect. The decrease in the phase spacing and sharp increase in the secondary phase content resulted in a dramatic increase in the corrosion rate of the ZM-1.0Ca alloy.

Keywords: Mg-Zn-Mn-Ca; Ca$_2$Mg$_6$Zn$_3$ phase; corrosion behaviors; SKPFM

Citation: Fu, J.; Du, W.; Liu, K.; Du, X.; Zhao, C.; Liang, H.; Mansoor, A.; Li, S.; Wang, Z. Effect of the Ca$_2$Mg$_6$Zn$_3$ Phase on the Corrosion Behavior of Biodegradable Mg-4.0Zn-0.2Mn-xCa Alloys in Hank's Solution. *Materials* **2022**, *15*, 2079. https://doi.org/10.3390/ma15062079

Academic Editor: Frank Czerwinski

Received: 17 January 2022
Accepted: 7 March 2022
Published: 11 March 2022

Publisher's Note: MDPI stays neutral with regard to jurisdictional claims in published maps and institutional affiliations.

Copyright: © 2022 by the authors. Licensee MDPI, Basel, Switzerland. This article is an open access article distributed under the terms and conditions of the Creative Commons Attribution (CC BY) license (https://creativecommons.org/licenses/by/4.0/).

1. Introduction

Magnesium (Mg) alloys have attracted great attention as promising biodegradable materials for orthopedic implants and cardiovascular interventional devices [1–5]. Compared with traditional metallic biomaterials, such as stainless steels and titanium alloys, Mg alloys reduce the stress-shielding risk [6]. Mg alloys usually degrade within a few weeks, and the produced Mg^{2+} ions can be advantageous for bone healing without toxicity [7]. However, the uncontrolled corrosion of Mg alloys may lead to unexpected mechanical failure before tissue recovery. This is a major problem for Mg alloys being used as clinical implants in a physiological environment with a high chloride and/or pH of 7.4–7.6 [7]. Therefore, controlling the corrosion of Mg alloys is becoming an urgent problem.

Alloying is one of the most effective methods to improve corrosion resistance and sustain the mechanical properties of Mg alloys [8,9]. Aluminum (Al) and rare earth elements are usually selected to modify the degradation rate of Mg alloys, but their adverse effects on biocompatibility with the human body have to be fully considered. For example, excessive Al usually causes nerve toxicity and restrains human body growth [10–14], while rare earth elements increase the risk of thrombosis [3]. In addition, alloying may result in precipitation of secondary phases and typically accelerate galvanic corrosion of Mg alloys, leading to the consequence of local corrosion and collapse of Mg implants. Thus, it is necessary to design and control secondary phases by choosing reasonable alloying elements, as well as heat treatments, to reduce their impacts on the corrosion rate of Mg alloys.

Manganese (Mn) is a beneficial element for improving the corrosion resistance of Mg alloys [15–18]. Studies have shown that Mn addition resulted in forming a MnO_2 film on the surface of Mg alloys, inhibiting the permeation of chloride ions and improving corrosion resistance [17,18]. Zinc (Zn) is one of the most essential elements for the physiological functions of the human body [2], but Zn accelerates the corrosion rate because of forming the Mg_xZn_y phase [19–21]. As reported by Song [22], the Mg_xZn_y phase acted as a microcathode and accelerated corrosion as its volume fraction increased. Calcium (Ca) is an indispensable element of human bone, which not only accelerates bone healing [23] but also enhances corrosion resistance and mechanical performance of the implant. When Ca is added to Mg-Zn alloys, the Mg_2Ca and $Ca_2Mg_6Zn_3$ phases are usually formed according to the atomic ratio of Zn/Ca [24–26]. Although previous studies have indicated that the $Ca_2Mg_6Zn_3$ phase has dominating effects on the corrosion rate of Mg-Zn-Ca alloys [27–30], its detailed corrosion behavior has not been entirely understood. In this work, we have focused on the effects of the $Ca_2Mg_6Zn_3$ phase with respect to its morphology, distribution and volume fractions on the corrosion behavior of Mg-4.0Zn-0.2Mn-xCa alloys in Hank's simulate solution, a solution similar to human body fluid, which is generally used for in vitro corrosion experiments. Its composition is given in Section 2.4. The corrosion mechanism is discussed.

2. Materials and Methods

2.1. Material Fabrication

The materials used in this work were cast in the laboratory. The as-cast Mg-4.0Zn-0.2Mn-xCa alloys (denoted as ZM-xCa alloys, x = 0.1, 0.3, 0.5 and 1.0 wt.%) were prepared using high-purity Mg (99.95 wt.%), high-purity Zn (99.99 wt.%), Mg-Mn and Mg-Ca master alloys procured from Hunan Research and Institute of Rare Earth and Material. Melting (Changsha, China) was conducted in an electrical resistance furnace under a mixed gas of N_2 and SF_6 (the ratio of N_2 and SF_6 was 100:1). The melt was kept at 750 °C for 20 min, then poured into a steel mold (preheated at 200 °C) at 720 °C and cooled to room temperature. The actual chemical compositions of the ZM-xCa alloys were examined by X-ray fluorescence (XRF, Magix-PW2403 with a test resolution of 0.0001), as shown in Table 1.

Table 1. Actual compositions of the ZM-xCa alloys (wt.%).

Alloys	Actual Compositions (wt.%)				
	Zn	Mn	Ca	Si	Mg
ZM-0.1Ca	4.13	0.20	0.12	<0.02	Bal.
ZM-0.3Ca	3.81	0.17	0.34	<0.02	Bal.
ZM-0.5Ca	3.78	0.18	0.56	<0.03	Bal.
ZM-1.0Ca	3.76	0.18	0.95	<0.02	Bal.

2.2. Microstructural Characterization

Microstructure observation was conducted using an optical microscope (AXIO IMAGER A2M) and a scanning electron microscope (SEM, HITACHI S3400N, HITACHI, Tokoy, Japan) equipped with energy dispersive X-ray spectroscopy (EDS). The nominal distance between secondary phases (phase spacing) and the volume fraction of precipitates were determined by Image-Pro Plus 6.0 software (Media Cybernetics, Houston, TX, USA), and 10 images were selected for data statistics at least. The samples for optical microscope observation were ground up to 5000# SiC grit papers (Federation of European Producers of Abrasives, FEPA standard) and polished with 0.5 μm diamond paste, etched by 5% HNO_3/C_2H_6O solution for 10~20 s. X-ray diffraction (XRD, D/MAX-3C, Rigaku, Tokyo, Japan) with Cu Kα radiation was used to analyze phases with a step of 0.02° in the scanning range of 10~90° at room temperature. The microstructure features of secondary phases were characterized by transmission electron microscopy (TEM, JEM-2100, JEOL, Tokyo, Japan) equipped with an Oxford energy spectrum system (EDX can give semi-quantitative analysis, Oxford Instruments, Abingdon, UK) at an accelerating voltage of 200 kV. The dwelling time was 100 ms, and the anode voltage was 15 mV, with a current of 10 mA. Three-dimensional (3D) distribution of secondary phases was analyzed by X-ray Microscope (XRM, Xradia 520 Versa, ZEISS, Oberkochen, Baden-Württemberg, Germany) with a high effective pixel size of 1.5 μm at a voltage of 80 kV and a power of 7 W. The specimen (cylindrical shape) with dimensions of Φ1.8 mm × 25 mm was prepared for the 3D morphology test. The Object Research System (ORS) Visual software (Object Research System, Montreal, QC, Canada) was used to analyze the experimental data obtained by X-ray Microscope to reconstruct the 3D tomography image of the secondary phase.

2.3. Local Volta Potential Measurement

The potentials (relative nobility potential) of secondary phases were measured by scanning kelvin probe force microscopy (SKPFM, Bruker Icon, Bruker, Billerica, MA, USA) in the tapping mode at room temperature (~25 °C) and a relative humidity of ~50%. A magnetic etched silicon probe was used to measure the relative Volta potential, with a bias potential of 5 V applied to the sample. The tip height is 100 nm, pixel resolution is 20 nm, and scan rate is 0.5 Hz. Prior to the SKPFM test, the samples were polished with diamond paste and alcohol. The samples were vacuum packed immediately after drying with cold air. The Volta potential differences were analyzed using NanoScope Analysis 2.0 software (Bruker, Billerica, MA, USA).

2.4. In Vitro Immersion Test

The immersion test was carried out in Hank's physiological solution (8.00 g/L NaCl, 0.40 g/L KCl, 0.14 g/L $CaCl_2$, 0.35 g/L $NaHCO_3$, 0.1 g/L $MgCl_2·6H_2O$, 0.06 g/L $MgSO_4·7H_2O$, 0.06 g/L KH_2PO_4 and 0.06 g/L $Na_2HPO_4·12H_2O$, pH = 7.4) at 37 ± 0.4 °C with different periods. The samples (10 × 10 × 10 mm^3) used for corrosion studies were cut from the cast block, and then ground up to 2000# SiC grit papers. Subsequently, the samples were cleaned ultrasonically with acetone and ethanol solutions, respectively. The ratio of sample surface area to solution volume was 1 cm^2:150 mL and the solution was renewed every two days. After immersion, the samples for SEM were ultrasonically cleaned with 200 g/L CrO_3 + 10 g/L $AgNO_3$ solution for 10 min to remove the surface corrosion products. The average corrosion rate (estimated by weight loss, V_C) was calculated according to the following equation [31,32]:

$$V_C = \frac{3650 \times (W_o - W_t)}{DAT} \quad (1)$$

where, V_C is the corrosion rate (mm/y), W_o is the sample weight before immersion (g), W_t is the sample weight after removing corrosion products (g), D is the density (g/cm^3), A is the sample surface area exposed to the solution (cm^2), and T is the immersion time (d).

2.5. Electrochemical Measurement

The samples were sealed in epoxy resin to expose only one surface (1 cm^2) for electrochemical measurement. The measurement was conducted by an electrochemical workstation (Autolab, Metrohm, Herisau, Appenzell Ausserrhoden, Switzerland) with a three-electrode system, in which the saturated calomel electrode (SCE), the platinum mesh and the sample were used as reference electrode, counter electrode and working electrode, respectively. The potentiodynamic polarization tests were performed in Hank's solution (about 200 mL) at a constant scanning rate of 0.5 mv/s, and the voltage range to open circuit potential (OCP) was ±300 mv, and the OCP monitored time was 120 s. The electrochemical parameters, such as corrosion potential (E_{corr}) and corrosion current density (i_{corr}) were obtained by Tafel extrapolation. The electrochemical impedance spectroscopy

(EIS) measurement was carried out in a frequency range from 100 kHz to 100 mHz with a perturbation of 10 mV, and the EIS data (Frequency, Z' and Z", about 220 data points) were fitted using ZView3.1 software (Scribner Associates, Southern Pines, NC, USA). All measurements were repeated at least three times to ensure the reproducibility of the results.

3. Results

3.1. Microstructure of ZM-xCa Alloys

Figure 1 shows the OM (a–d) and SEM (a_1–d_1) micrographs of the ZM-xCa alloys. As shown in Figure 1, the dendritic and spherical phases are mainly distributed along grain boundaries and within grains, respectively. Although the distribution of the secondary phases hardly changed, the grain size decreased with an increase in the Ca amount. Table 2 lists the microstructure characteristics of the ZM-xCa alloys.

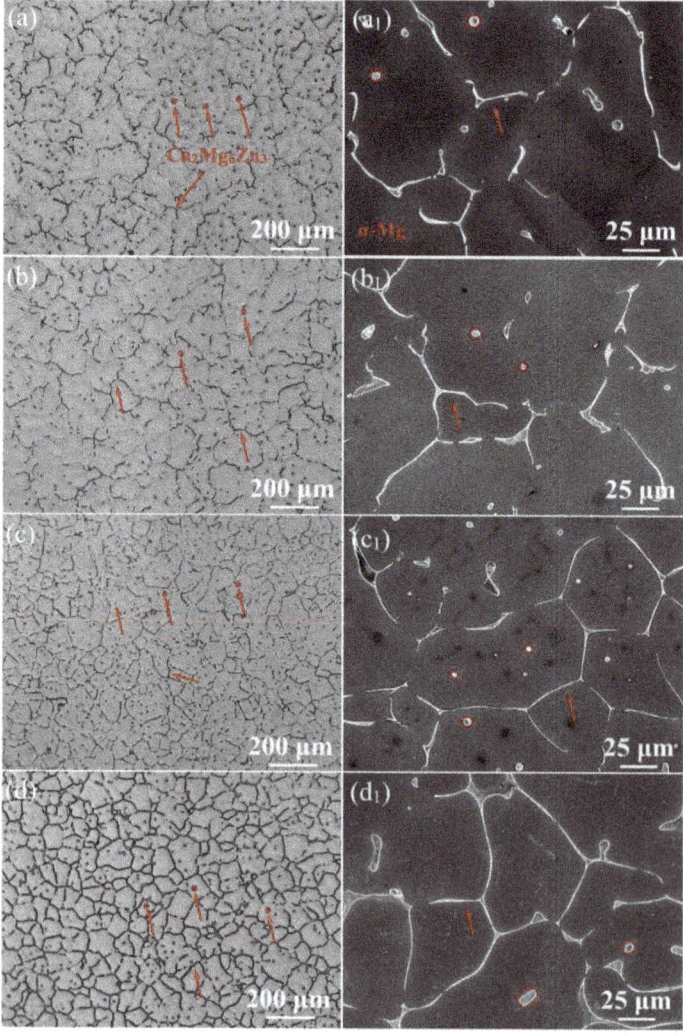

Figure 1. OM (**a–d**) and SEM (a_1–d_1) micrographs of the as-cast ZM-0.1Ca (**a**,a_1), ZM-0.3Ca (**b**,b_1), ZM-0.5Ca (**c**,c_1) and ZM-1.0Ca (**d**,d_1) alloys.

Table 2. Microstructure characteristics of the ZM-xCa alloys (wt.%).

Alloys	Volume Fraction of Secondary Phase (%)	Secondary Phase Spacing (μm)
ZM-0.1Ca	2.5 ± 0.2	43.4 ± 2.0
ZM-0.3Ca	3.0 ± 0.2	39.3 ± 2.1
ZM-0.5Ca	4.2 ± 0.3	32.5 ± 1.5
ZM-1.0Ca	7.6 ± 0.4	30.1 ± 1.2

This indicates that the total volume fraction of the secondary phases increased from 2.5% to 7.6%, while spacing declined from 43 μm to 30 μm when the added Ca amount increased from 0.1 wt.% to 1.0 wt.%.

Figure 2 shows the XRD patterns of the ZM-xCa alloys. Only α-Mg and $Ca_2Mg_6Zn_3$ peaks are present in all ZM-xCa alloys. In addition, the diffraction intensity of the $Ca_2Mg_6Zn_3$ peak increased with an increase in the Ca amount, indicating that Ca promoted the eutectic precipitation of the $Ca_2Mg_6Zn_3$ phase. Figure 3 shows the TEM images and EDS results of the ZM-0.3Ca and ZM-0.5Ca alloys. The phases in these two alloys were located at grain boundaries and had a strip morphology. According to the EDS results, these secondary phases are inferred to be $Ca_2Mg_6Zn_3$, showing a hexagonal crystal structure [33–36].

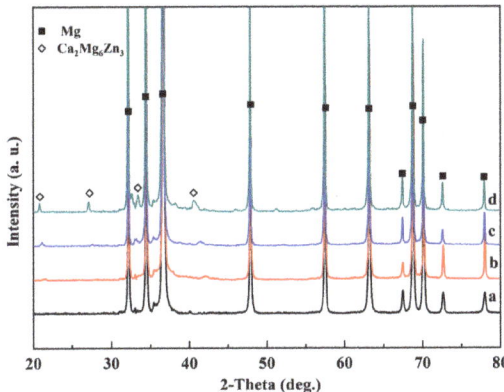

Figure 2. XRD patterns of ZM-xCa alloys: (**a**) ZM-0.1Ca, (**b**) ZM-0.3Ca, (**c**) ZM-0.5Ca and (**d**) ZM-1.0Ca.

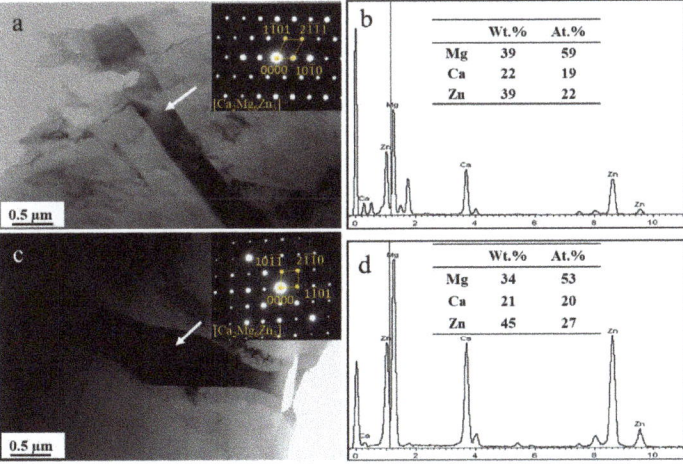

Figure 3. TEM images and EDS results of the secondary phases: (**a,b**) ZM-0.3Ca and (**c,d**) ZM-0.5Ca alloys.

3.2. Electrochemical Properties of the ZM-xCa Alloys

Two typical $Ca_2Mg_6Zn_3$ phases, namely the spherical (Figure 4a) and dendritic (Figure 4b) phases in the ZM-0.3Ca alloy, were chosen to conduct SKPFM analysis. The Volta potential differences between $Ca_2Mg_6Zn_3$ phases and Mg matrix are revealed in Figure 4c–f.

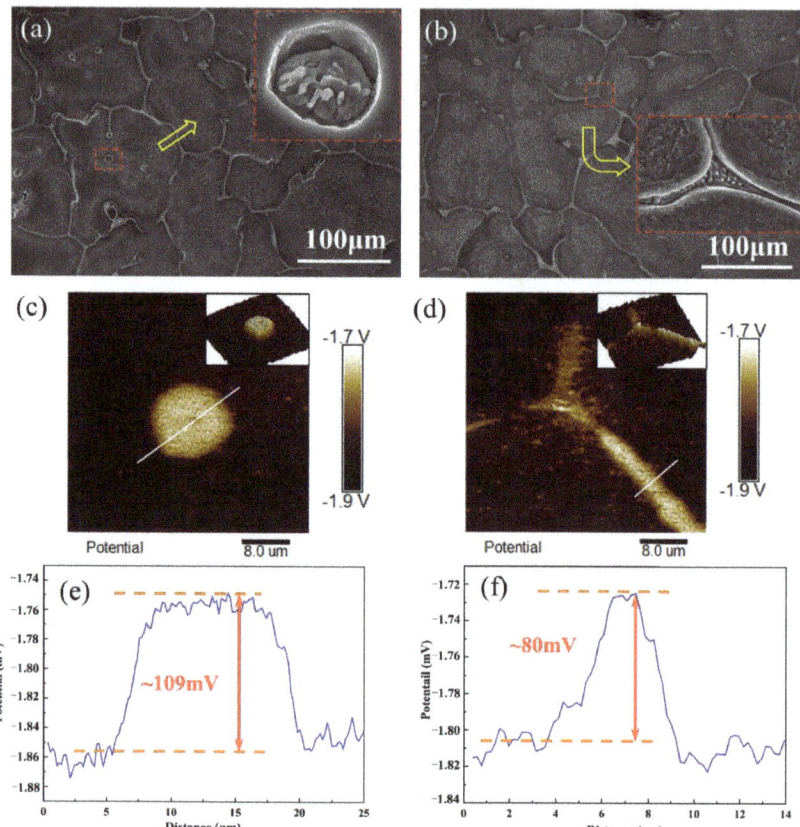

Figure 4. SEM images and SKPFM analysis of the ZM-0.3Ca alloy: (**a**) SEM images of the spherical secondary phase and (**b**) the dendritic secondary phase; (**c**) SKPFM maps of the spherical secondary phase and (**d**) the dendritic secondary phase; (**e**) the Volta potential profile of line as indicated in (**c**); (**f**) the Volta potential profile of line as indicated in (**d**).

From Figure 4c,d, it can be determined whether the spherical or the dendritic phase is brighter than the Mg matrix. The SKPFM work function mode supports the idea that the light color demonstrates more positive potential than the dark one, i.e., the light color represents the cathodic area, whereas the dark one stands for anodic one [37,38]. According to the line-profile analyses shown in Figure 4e,f, the spherical phase exhibited a relatively higher Volta potential difference (about +109 mV) than the dendritic phase (about +80 mV). These results imply that the spherical phase has a stronger acceleration effect than micro-galvanic corrosion.

The potentiodynamic polarization curves of the ZM-xCa alloys in Hank's solution are presented in Figure 5a. The values of electrochemical parameters, i.e., corrosion potential (E_{corr}), corrosion current density (i_{corr}) and breakdown potential (E_b) are summarized in Table 3. It can be found that the E_{corr} moved in the negative direction, but the i_{corr}

consistently increased with an increase in the Ca amount, and the ZM-0.1Ca alloy displayed the lowest i_{corr} value of 5.90×10^{-7} A/cm^2. In the potentiodynamic polarization test, the corrosion potential is a thermodynamic parameter that reflects the probability of the corrosion tendency, while the corrosion current density is a kinetic parameter that signifies the corrosion rate [39]. The most positive corrosion potential, as well as the lowest corrosion current density of the ZM-0.1Ca alloy, indicates that it has the best corrosion resistant property. Moreover, the ZM-0.1Ca alloy has the most positive breakdown potential (Eb) of −1.450 V, which indicates that the corrosion reaction is more difficult to proceed because of the formed corrosion film [40].

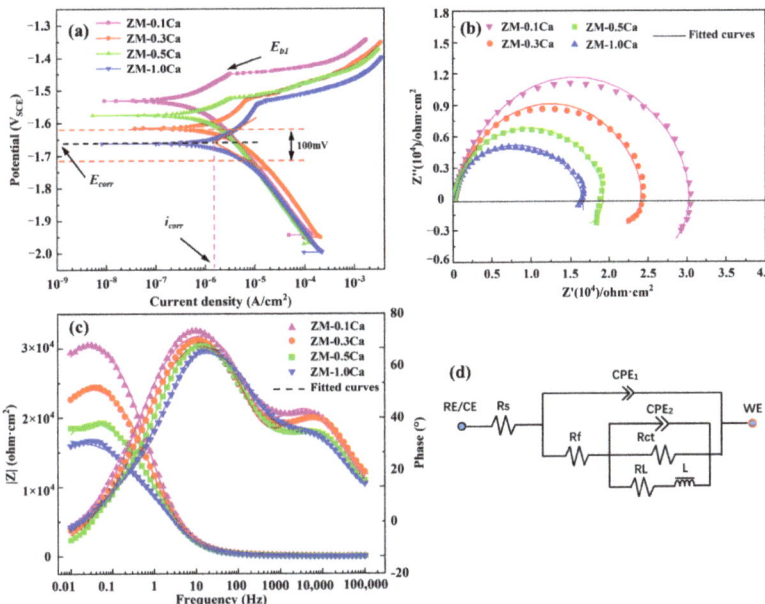

Figure 5. Potentiodynamic polarization curves (a) and electrochemical impedance spectra of the ZM-xCa alloys: (b) Nyquist plots, (c) Bode impedance and Bode phase angle plots, and (d) Equivalent circuit.

Table 3. Electrochemical parameters of the ZM-xCa alloys.

Alloys	E_{corr} (V)	i_{corr} (A/cm^2)	E_b (V)
ZM-0.1Ca	−1.530	5.90×10^{-7}	−1.450
ZM-0.3Ca	−1.575	6.63×10^{-7}	−1.524
ZM-0.5Ca	−1.617	6.77×10^{-7}	−1.527
ZM-1.0Ca	−1.662	1.31×10^{-6}	−1.542

Figure 5b–d indicates the EIS results of the ZM-xCa alloys. As shown in Figure 5b, the Nyquist plots of the four alloys are composed of capacitance loops at high frequency and an inductive loop at low frequency. All of the Nyquist plots are similar in shape except for the diameter of the loop. This result demonstrates that the corrosion mechanism of the four alloys is similar, but their corrosion rates are different. According to the Nyquist plots, the corrosion rate ranks as ZM-0.1Ca < ZM-0.3Ca < ZM-0.5Ca < ZM-1.0Ca. In addition, the inductive loop at a low frequency implies a breakdown of the protective corrosion-products film [41]. Figure 5c shows the impedance modulus (|Z|), as well as the Bode phase angle plots of the ZM-xCa alloys. The largest impedance modulus (|Z|) of the ZM-0.1Ca alloy represents the best corrosion protection resulting from the corrosion-product film. The maximum crest of the Bode phase angle plot also reflects the difficulty of charge transfer of

the ZM-0.1Ca alloy. Figure 5d indicates the fitted equivalent circuit using the EIS data. R_s, R_f and R_{ct} are solution resistance, film resistance and charge transfer resistance, respectively. CPE_1 represents the constant phase element of corrosion product film, CPE_2 represents the double layer capacitance. L and R_L are the inductance and inductance resistance. The detailed fitting results are listed in Table 4.

Table 4. The fitting results of the EIS spectra of ZM-xCa alloys.

Alloys	R_s (Ω cm^2)	R_f (Ω cm^2)	CPE_1 (10^{-6} F/cm^2)	n_1	CPE_2 (10^{-6} F/cm^2)	n_2	R_{ct} (kΩ cm^2)	R_L (kΩ cm^2)	L (H cm^2)	R_P (kΩ cm^2)
ZM-0.1Ca	25.48	281.31	7.60	0.73	6.31	0.90	31.29	53.97	2.26×10^6	20.11
ZM-0.3Ca	22.41	278.43	9.13	0.70	6.66	0.89	25.17	70.05	1.61×10^6	18.80
ZM-0.5Ca	22.40	230.21	11.71	0.69	8.83	0.85	45.69	20.25	1.12×10^6	14.28
ZM-1.0Ca	21.57	206.30	19.08	0.66	5.16	0.88	10.80	16.90	2.30×10^6	6.86

The polarization resistance (R_p), which is inversely proportional to the corrosion rate, is an important parameter for evaluating the corrosion performance of an alloy. Its value was calculated using the following equation:

$$R_P = R_S + R_f + \frac{R_{ct} R_L}{R_{ct} + R_L} \quad (2)$$

The calculated R_p of the ZM-xCa alloys is also listed in Table 4. The largest R_p of the ZM-0.1Ca alloy demonstrated its best corrosion resistance, which is consistent with the polarization result.

3.3. Corrosion Behaviors of ZM-xCa Alloys

Figure 6 shows the corrosion rates of the ZM-xCa alloys by weight loss after immersion in Hank's solution for 7 days and 14 days, respectively.

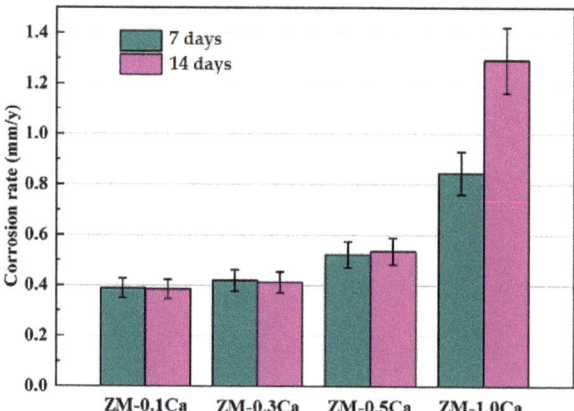

Figure 6. Corrosion rates of the ZM-xCa alloys obtained by weight loss after immersion in Hank's solution for different times.

The corrosion rate increases with an increase in the added Ca amount for neither 7-day nor 14-day immersion. The corrosion rate increased slowly when the added Ca amount was less than 0.5 wt.%; however, it sharply increased for the ZM-1.0Ca alloy. In particular, the 14-day corrosion rate of the ZM-1.0Ca alloy is about three times that of the ZM-0.1Ca alloy. These results demonstrate that excessive Ca addition destructs the corrosion resistance of the ZM-xCa alloys.

The surface morphologies after removal of corrosion products of the ZM-xCa alloys with various immersion times are displayed in Figure 7. After 7-day immersion, a shallow corrosion area extended with an increase in the added Ca amount, as shown in Figure 7a–c,e. On the other hand, corrosion morphology also changes significantly with corrosion time. In the case of the ZM-1.0Ca alloy, plenty of tiny corrosion pits appeared after 3-day immersion (Figure 7d), and an almost whole corroded morphology was presented after 7-day immersion (Figure 7e). The corrosion pits became corrosion cavities with an extension of the corrosion time. The corrosion cavity became deeper and wider and reached about 554 μm after 14-day immersion (Figure 7d$_1$–f$_1$).

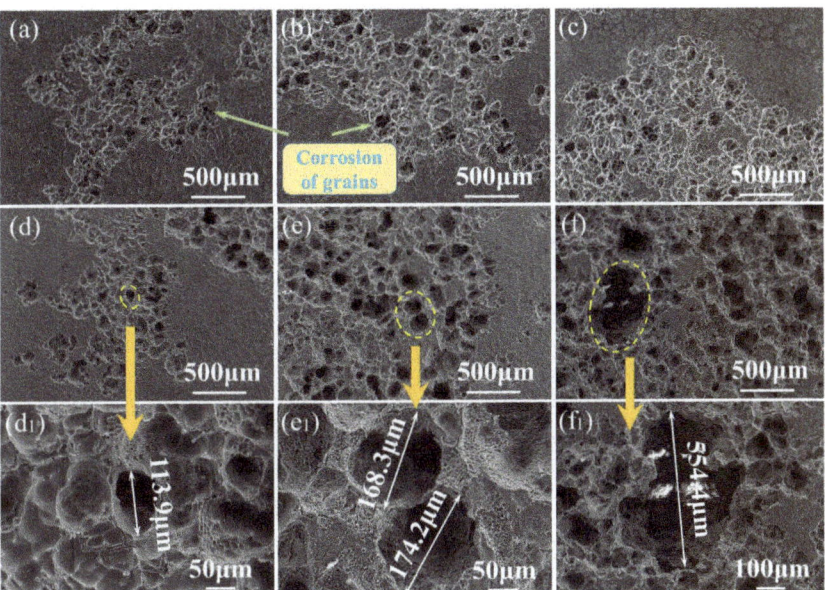

Figure 7. Surface morphologies after removal of the corrosion products with various immersion times in Hank's solution of the ZM-xCa alloys: (**a**–**c**) ZM-0.1Ca, ZM-0.3Ca and ZM-0.5Ca alloys immersed for 7-day; (**d,d$_1$**), (**e,e$_1$**) and (**f,f$_1$**) ZM-1.0Ca alloy immersed for 3-day, 7-day and 14-day, respectively.

4. Discussion

The Ca$_2$Mg$_6$Zn$_3$ phase plays a significant role in the corrosion process of magnesium alloys [42,43]. In the present study, the corrosion behavior of the ZM-xCa alloys is investigated by altering the Ca$_2$Mg$_6$Zn$_3$ phase morphology, distribution and volume fraction. Figure 8 shows the 3D topographical characteristics of the Ca$_2$Mg$_6$Zn$_3$ phase in the ZM-0.3Ca alloy, giving an intuitive distribution of the secondary phase. It was found that there are two kinds of morphologies of the Ca$_2$Mg$_6$Zn$_3$ phase, one spherical and the other dendritic. In addition, the dendritic (marked with red color) Ca$_2$Mg$_6$Zn$_3$ phase precipitated at grain boundaries and connected with each other, forming a continuous network, while the spherical (marked with violet color) Ca$_2$Mg$_6$Zn$_3$ phase was distributed within grains. These results are consistent with the optical images shown in Figure 1.

The Volta potential in air is widely used in the corrosion field to assess the corrosion tendency of Mg alloys [42–44]. According to the SKPFM results shown in Figure 4, the Volta potential of either the dendritic or the spherical Ca$_2$Mg$_6$Zn$_3$ phase is higher than that of the Mg matrix; hence, not only the dendritic but also the spherical Ca$_2$Mg$_6$Zn$_3$ phase acts as a cathode, while the Mg matrix acts as an anode in micro-galvanic corrosion. As a result, the Mg matrix dissolves preferentially during the immersion period. Therefore, the total amount of the Ca$_2$Mg$_6$Zn$_3$ phase, as well as its phase spacing, plays a controlling role in the corrosion process; that is, the higher the volume fraction and/or the smaller the phase spacing, the higher the corrosion rate. Figure 9 shows the relationship between the corrosion rate, the volume fraction and the spacing of the secondary phase in the ZM-xCa alloys. With the increase in the volume fraction and decrease in the spacing of the secondary phase, the corrosion rate of the ZM-xCa alloys after 14-day immersion increases. The increase in the total secondary phase creates more micro-galvanic positions, and the shortening phase spacing decreases the distance between the corrosion pits. These two aspects simultaneously promoted the corrosion rate of the ZM-xCa alloys.

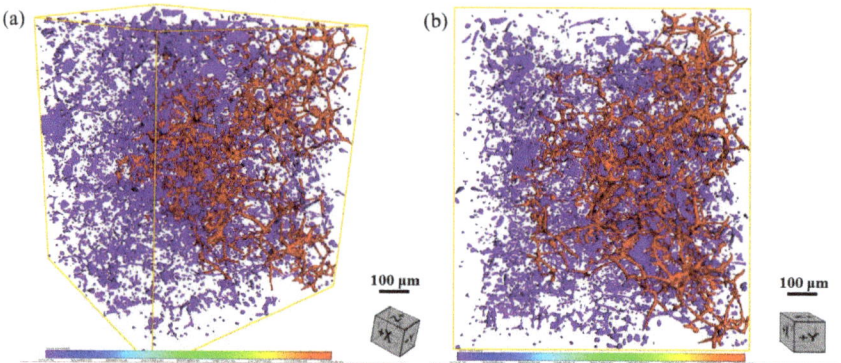

Figure 8. Three-dimensional topographical characteristics of the $Ca_2Mg_6Zn_3$ secondary phase in the ZM-0.3Ca alloy: (**a**) perspective view, (**b**) sectional view of the perpendicular Y axle.

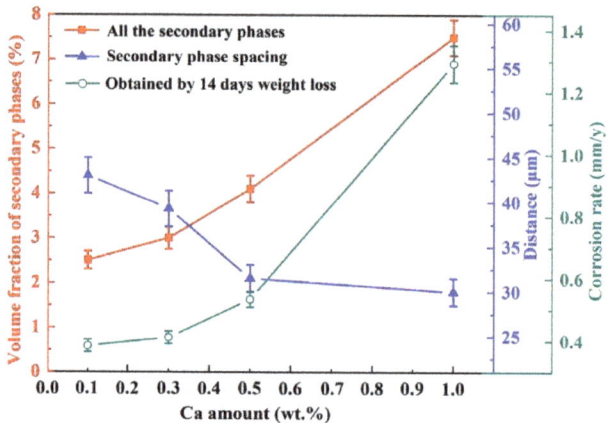

Figure 9. Relationship between corrosion rate, secondary phase spacing and volume fraction of the ZM-xCa alloys.

Because the SKPFM results show that the Volta potential of the spherical $Ca_2Mg_6Zn_3$ phase is higher than that of the dendritic one, the spherical phase is electrochemically preferred as a cathode to accelerate Mg matrix (anode) dissolution in the micro-galvanic couple.

We discussed the ion exchange process and formation of the corrosion products during the corrosion process of Mg alloy in our previous studies [3,31]. We do not discuss this in the present investigation. The present research focuses on the effect of secondary phase characteristics on the corrosion behavior of Mg alloys. Figure 10 intuitively illustrates a schematic view of the corrosion behavior of the ZM-0.1Ca and ZM-1.0 alloys. The corrosion process occurs preferentially around spherical phases in stage I (as shown in Figure 10a,d). Because of the dissolution of the Mg matrix around the spherical phase, it falls off and forms corrosion pits in stage II. If the phase spacing is large, for example, in the ZM-0.1Ca alloy, the corrosion process is difficult to proceed to another phase; hence, small corrosion pits are formed (as shown in Figure 10b). As immersion continues, the corrosion process proceeds into grains, and the corrosion process occurs around the dendritic phase distributing at grain boundaries. This corrosion is different from that in grains. Instead of forming corrosion pits, a corrosion "channel" along the dendritic phase is formed, which induces the corrosion process to proceed into the interior grains, as shown in Figure 10b,d. For the alloy with a high Ca content, corrosion penetrates speedily and continuously along grain boundaries. Because of falling off these dendritic phases and dissolution of grains, large corrosion cavities are formed in stage III (see Figure 10f), resulting in a rapid corrosion rate. However, in the case of Zn-0.1Ca alloy, the corrosion pits caused by pitting corrosion are difficult to connect; hence, independent small-scale corrosion pits are generated in the Zn-0.1Ca alloy (as shown in Figure 10c), resulting in slow mass loss.

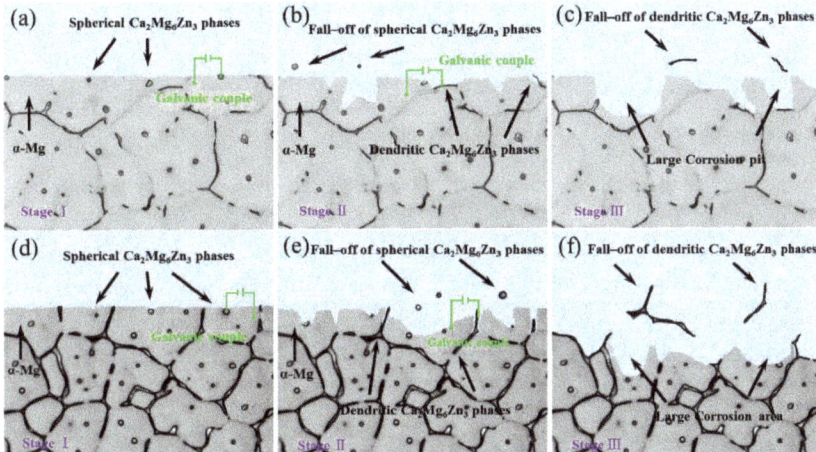

Figure 10. Schematic illustration of corrosion process of the investigated alloys in Hank's solution: (**a**–**c**) ZM-0.1Ca alloy and (**d**–**f**) ZM-1.0Ca alloy.

5. Conclusions

1. The cast ZM-xCa (0.1, 0.3, 0.5 and 1.0 wt.%) alloys contain the intragranular spherical $Ca_2Mg_6Zn_3$ phase and the dendritic one at grain boundaries. The volume fraction of the $Ca_2Mg_6Zn_3$ phase gradually increased from 2.5% to 7.6%, while its spacing declined monotonically from 43.0 μm to 30.0 μm with an increase in the added Ca amount from 0.1 wt.% to 1.0 wt.%.
2. The Volta potential of the spherical $Ca_2Mg_6Zn_3$ phase (+109 mV) was higher than that of the dendritic phase (+80 mV); hence, the spherical phase was electrochemically preferred as a cathode to accelerate Mg matrix (anode) dissolution in the micro-galvanic couple.
3. The corrosion rate obtained by weight loss increased slightly with increasing Ca content from 0.1 wt.% to 0.5 wt.% because of the enhanced micro-galvanic corrosion effect. The decrease in the phase spacing and sharp increase in the secondary phase content resulted in a dramatic increase in the corrosion rate of the ZM-1.0Ca alloy.

Author Contributions: Conceptualization, J.F. and W.D.; methodology, J.F. and W.D.; software, J.F.; validation, W.D., K.L. and H.L.; formal analysis, J.F. and S.L.; investigation, J.F., A.M. and X.D.; resources, C.Z. and Z.W.; J.F. data curation; J.F. writing—original draft preparation; W.D., K.L. and H.L. writing—review and editing. All authors have read and agreed to the published version of the manuscript.

Funding: This research was funded by the Beijing Natural Science Foundation (No. 2192006) and the National Natural Science Foundation of China (No. 51801004; No. 52001015).

Institutional Review Board Statement: Not applicable.

Informed Consent Statement: Not applicable.

Data Availability Statement: All data are available within the manuscript.

Conflicts of Interest: The authors declare no conflict of interest.

References

1. Birbilis, N.; Easton, M.A.; Sudholz, A.D.; Zhu, S.M.; Gibson, M.A. On the corrosion of binary magnesium-rate earth alloys. *Corros. Sci.* **2009**, *51*, 683–689. [CrossRef]
2. Zhang, S.X.; Zhang, X.N.; Zhao, C.L.; Li, J.N.; Song, Y.; Xie, C.Y.; Tao, H.R.; Zhang, Y.; He, Y.H.; Jiang, Y.; et al. Research on an Mg–Zn alloy as a degradable biomaterial. *Acta Biomater.* **2010**, *6*, 626–640. [CrossRef]
3. Du, W.B.; Liu, K.; Ma, K.; Wang, Z.H.; Li, S.B. Effects of trace Ca/Sn addition on corrosion behaviors of biodegradable Mg-4Zn-0.2Mn alloy. *J. Magnes. Alloy* **2018**, *6*, 1–14. [CrossRef]
4. Munir, K.; Lin, J.X.; Wen, C.E.; Wright, P.F.A.; Li, Y.C. Mechanical, corrosion, and biocompatibility properties of Mg-Zr-Sr-Sc alloys for biodegradable implant applications. *Acta Biomater.* **2020**, *102*, 493–507. [CrossRef]
5. Witte, F. The history of biodegradable magnesium implants. *Acta Biomater.* **2010**, *6*, 1680–1692. [CrossRef]
6. Lu, Y.; Bradshaw, A.R.; Chiu, Y.L.; Jones, I.P. Effects of secondary phase and grain size on the corrosion of biodegradable Mg-Zn-Ca alloys. *Mater. Sci. Eng. C* **2015**, *48*, 480–486. [CrossRef]
7. Witte, F.; Kaese, V.; Haferkamp, H.; Switzer, E.; Meyer-Lingenberg, A.; Wirth, C.J.; Windhagen, H. In vivo corrosion of four magnesium alloys and the associated bone response. *Biomaterials* **2005**, *26*, 3557–3563. [CrossRef] [PubMed]
8. Zhao, M.C.; Liu, M.; Song, G.L.; Atrens, A. Influence of the β-phase morphology on the corrosion of the Mg alloy AZ91. *Corros. Sci.* **2008**, *50*, 1939–1953. [CrossRef]
9. Atrens, A.; Song, G.L.; Liu, M.; Shi, Z.M.; Cao, F.Y.; Dargusch, M.S. Review of recent developments in the field of magnesium corrosion. *Adv. Eng. Mater.* **2015**, *17*, 400–453. [CrossRef]
10. Cai, C.H.; Alvens, M.M.; Song, R.B.; Wang, Y.J.; Li, J.Y.; Montemor, M.F. Non-destructive corrosion study on a magnesium alloy with mechanical properties tailored for biodegradable cardiovascular stent applications. *J. Mater. Sci. Technol.* **2021**, *66*, 128–138. [CrossRef]
11. Zengin, H.; Turen, Y. Effect of Y addition on microstructure and corrosion behavior of extruded Mg–Zn–Nd–Zr alloy. *J. Magnes. Alloy* **2020**, *8*, 640–653. [CrossRef]
12. Jin, S.; Zhang, D.; Lu, X.P.; Zhang, Y.; Tan, L.L.; Liu, Y.; Wang, Q. Mechanical properties, biodegradability and cytocompatibility of biodegradable Mg-Zn-Zr-Nd/Y alloys. *J. Mater. Sci. Technol.* **2020**, *47*, 190–201. [CrossRef]
13. Zhang, E.L.; Yang, L. Microstructure, mechanical properties and bio-corrosion properties of Mg-Zn-Mn-Ca alloy for biomedical application. *Mater. Sci. Eng. A* **2008**, *497*, 111–118. [CrossRef]
14. Ding, Y.F.; Wen, C.E.; Hodgson, P.; Li, Y.C. Effects of alloying elements on the corrosion behavior and biocompatibility of biodegradable magnesium alloys: A review. *J. Mater. Chem. B* **2014**, *2*, 1912–1933. [CrossRef] [PubMed]
15. Metalnikov, P.; Ben-Hamu, G.; Templeman, Y.; Shin, K.S.; Meshi, L. The relation between Mn additions, microstructure and corrosion behavior of new wrought Mg-5Al alloys. *Mater. Charact.* **2018**, *145*, 101–115. [CrossRef]
16. Wang, J.F.; Huang, S.; Li, Y.; Wei, Y.Y.; Xi, X.F.; Cai, K.Y. Microstructure, mechanical and bio-corrosion properties of Mn-doped Mg–Zn–Ca bulk metallic glass composites. *Mater. Sci. Eng. C* **2013**, *33*, 3832–3838. [CrossRef]
17. Cho, D.H.; Lee, B.W.; Park, J.Y.; Cho, K.M.; Park, I.M. Effect of Mn addition on corrosion properties of biodegradable Mg-4Zn-0.5Ca-xMn alloys. *J. Alloys Compd.* **2017**, *695*, 1166–1174. [CrossRef]
18. Bazhenov, V.E.; Li, A.V.; Komissarov, A.A.; Koltygin, A.V.; Tavolzhanskii, S.A.; Bautin, V.A.; Voropaeva, O.O.; Mukhametshina, A.M.; Tokar, A.A. Microstructure and mechanical and corrosion properties of hot-extruded Mg-Zn-Ca-(Mn) biodegradable alloys. *J. Magnes. Alloy* **2020**, *9*, 1428–1442. [CrossRef]
19. Cai, S.H.; Lei, T.; Li, N.F.; Feng, F.F. Effects of Zn on microstructure, mechanical properties and corrosion behavior of Mg-Zn alloys. *Mater. Sci. Eng. C* **2012**, *32*, 2570–2577. [CrossRef]
20. Li, H.; Peng, Q.M.; Li, X.J.; Li, K.; Han, Z.S.; Fang, D.Q. Microstructures, mechanical and cytocompatibility of degradable Mg-Zn based orthopedic biomaterials. *Mater. Des.* **2014**, *58*, 43–51. [CrossRef]
21. Zhang, E.L.; Yang, L.; Xu, J.W.; Chen, H.Y. Microstructure, mechanical properties and bio-corrosion properties of Mg-Si(-Ca, Zn) alloy for biomedical application. *Acta Biomater.* **2010**, *6*, 1756–1762. [CrossRef]
22. Song, Y.W.; Han, E.H.; Shan, D.Y.; Yim, C.D.; You, B.Y. The effect of Zn concentration on the corrosion behavior of Mg-xZn alloys. *Corros. Sci.* **2012**, *65*, 322–330. [CrossRef]
23. Zander, D.; Zumdick, N.A. Influence of Ca and Zn on the microstructure and corrosion of biodegradable Mg–Ca–Zn alloys. *Corros. Sci.* **2015**, *93*, 222–233. [CrossRef]

24. Erdmann, N.; Angrisani, N.; Reifenrath, J.; Lucas, A.; Thorey, F.; Bormann, D.; Meyer-Lindenberg, A. Biomechanical testing and degradation analysis of MgCa0.8 alloy screws: A comparative in vivo study in rabbits. *Acta Biomater.* **2011**, *7*, 1421–1428. [CrossRef]
25. Li, Z.J.; Gu, X.N.; Lou, S.Q.; Zheng, Y.F. The development of binary Mg-Ca alloys for use as biodegradable materials within bone. *Biomaterials* **2008**, *29*, 1329–1344. [CrossRef] [PubMed]
26. Kim, W.C.; Kim, J.G.; Lee, J.Y.; Seok, H.K. Influence of Ca on the corrosion properties of magnesium for biomaterials. *Mater. Lett.* **2008**, *62*, 4146–4148. [CrossRef]
27. Bakhsheshi-Rad, H.R.; Abdul-Kadir, M.R.; Idris, M.H.; Farahany, S. Relationship between the corrosion behavior and the thermal characteristics and microstructure of Mg-0.5Ca-xZn alloys. *Corros. Sci.* **2012**, *64*, 184–197. [CrossRef]
28. Asadi, J.; Korojy, B.; Hosseini, A.; Alishahi, M. Effect of cell structure on mechanical and bio-corrosion behavior of biodegradable Mg-Zn-Ca foam. *Mater. Today Commun.* **2021**, *28*, 102715. [CrossRef]
29. Gong, C.W.; He, X.Z.; Fang, D.Q.; Liu, B.S.; Yan, X. Effect of second phases on discharge properties and corrosion behaviors of the Mg-Ca-Zn anodes for primary Mg-air batteries. *J. Alloys Compd.* **2021**, *861*, 158493. [CrossRef]
30. Meng, X.; Jiang, Z.T.; Zhu, S.J.; Guan, S.K. Effects of Sr addition on microstructure, mechanical and corrosion properties of biodegradable Mg–Zn–Ca alloy. *J. Alloys Compd.* **2020**, *838*, 155611. [CrossRef]
31. Cheng, Y.F.; Du, W.B.; Liu, K.; Fu, J.J.; Wang, Z.H.; Li, S.B.; Fu, J.L. Mechanical properties and corrosion behaviors of Mg-4Zn-0.2Mn-0.2Ca alloy after long term in vitro degradation. *Trans. Nonferrous Met. Soc. China* **2020**, *30*, 363–372. [CrossRef]
32. Liu, J.; Yang, L.X.; Zhang, C.Y.; Zhang, B.; Zhang, T.; Li, Y.; Wu, K.M.; Wang, F.H. Significantly improved corrosion resistance of Mg-15Gd-2Zn-0.39Zr alloys: Effect of heat-treatment. *J. Mater. Sci. Technol.* **2019**, *35*, 1644–1654. [CrossRef]
33. Duley, P.; Sanyal, S.; Bandyopadhyay, T.K.; Mandal, S. Homogenization-induced age-hardening behavior and room temperature mechanical properties of Mg-4Zn-0.5Ca-0.16Mn (wt%) alloy. *Mater. Des.* **2019**, *164*, 107554. [CrossRef]
34. Zhang, Y.; Li, J.X.; Li, J.Y. Effects of calcium addition on phase characteristics and corrosion behaviors of Mg-2Zn-0.2Mn-xCa in simulated body fluid. *J. Alloys Compd.* **2017**, *728*, 37–46. [CrossRef]
35. Jiang, M.G.; Xu, C.; Nakata, T.; Yan, H.; Chen, R.S.; Kamado, S. Development of dilute Mg-Zn-Ca-Mn alloy with high performance via extrusion. *J. Alloys Compd.* **2016**, *668*, 13–21. [CrossRef]
36. Liu, C.Q.; Chen, X.H.; Chen, J.; Atrens, A.; Pan, F.S. The effects of Ca and Mn on the microstructure, texture and mechanical properties of Mg-4Zn alloy. *J. Magnes. Alloy.* **2020**, *9*, 1084–1097. [CrossRef]
37. Liu, J.H.; Song, Y.W.; Chen, J.C.; Chen, P.; Shan, D.Y.; Han, E.H. The Special Role of Anodic Second Phases in the Micro-galvanic Corrosion of EW75 Mg Alloy. *Electrochim. Acta* **2016**, *189*, 190–195. [CrossRef]
38. Cai, C.H.; Song, R.B.; Wang, L.X.; Li, J.Y. Effect of anodic T phase on surface micro-galvanic corrosion of biodegradable Mg-Zn-Zr-Nd alloys. *Appl. Surf. Sci.* **2018**, *462*, 243–254. [CrossRef]
39. Gui, Z.Z.; Kang, Z.X.; Li, Y.Y. Corrosion mechanism of the as-cast and as-extruded biodegradable Mg-3.0Gd-2.7Zn-0.4Zr-0.1Mn alloys. *Mater. Sci. Eng. C* **2019**, *96*, 831–840. [CrossRef]
40. Zhang, T.; Meng, G.Z.; Shao, Y.W.; Cui, Z.Y.; Wang, F.H. Corrosion of hot extrusion AZ91 magnesium alloy. Part II: Effect of rare earth element neodymium (Nd) on the corrosion behavior of extruded alloy. *Corros. Sci.* **2011**, *53*, 2934–2942. [CrossRef]
41. Bakhsheshi-Rad, H.R.; Idris, M.H.; Abdul-Kadir, M.R.; Ourdjini, A.; Medraj, M.; Daroonparvar, M.; Hamzah, E. Mechanical and bio-corrosion properties of quaternary Mg–Ca–Mn–Zn alloys compared with binary Mg–Ca alloys. *Mater. Des.* **2014**, *53*, 283–292. [CrossRef]
42. Pan, H.; Pang, K.; Cui, F.Z.; Ge, F.; Man, C.; Wang, X.; Cui, Z.Y. Effect of alloyed Sr on the microstructure and corrosion behavior of biodegradable Mg-Zn-Mn alloy in Hanks' solution. *Corros. Sci.* **2019**, *157*, 420–437. [CrossRef]
43. Yin, S.Q.; Duan, W.C.; Liu, W.H.; Wu, L.; Yu, J.M.; Zhao, Z.L.; Liu, M.; Wang, P.; Cui, J.Z.; Zhang, Z.Q. Influence of specific second phases on corrosion behaviors of Mg-Zn-Gd-Zr alloys. *Corros. Sci.* **2020**, *166*, 108419. [CrossRef]
44. Wei, L.Y.; Li, J.Y.; Zhang, Y.; Lai, H.Y. Effects of Zn content on microstructure, mechanical and degradation behaviors of Mg-xZn-0.2Ca-0.1Mn alloys. *Mater. Chem. Phys.* **2020**, *241*, 122441. [CrossRef]

Article

Ignition and Combustion Characteristic of B·Mg Alloy Powders

Yusong Ma, Kaichuang Zhang, Shizhou Ma, Jinyan He, Xiqiang Gai and Xinggao Zhang *

State Key Laboratory of NBC Protection for Civilian, Beijing 102205, China; mayusong111@163.com (Y.M.); 18333109527@163.com (K.Z.); 13910011793@139.com (S.M.); hejinyanfhy@163.com (J.H.); gaixq@126.com (X.G.)
* Correspondence: xinggaozhang@aliyun.com; Tel.: +86-13910664806

Abstract: Boron and its alloys have been explored a lot and it is expected that they can replace pure aluminum powder in the energetic formulation of active materials. MgB_2 compounds were prepared and characterized by a combination of mechanical alloying and heat treatment. The ignition and combustion of boron–magnesium alloys were studied with the ignition wire method and laser ignition infrared temperature measurement. The results show that MgB_2 has good ignition characteristics with maximum ignition temperatures obtained by the two various methods of 1292 K and 1293 K, respectively. Compared with boron, the ignition temperature of MgB_2 is greatly reduced after alloying. The ignition reaction of MgB_2 mainly occurs on the surface and the ignition process has two stages. In the initial stage of ignition, the large flame morphology and combustion state are close to the combustion with gaseous Mg, whereas the subsequent combustion process is close to the combustion process of B. Compared with boron, the ignition temperature of MgB_2 is greatly reduced which suggests that MgB_2 may be used in gunpowder, propellant, explosives, and pyrotechnics due to its improved ignition performance.

Keywords: boride; MgB_2; mechanical alloying; ignition; combustion; flame temperature

Citation: Ma, Y.; Zhang, K.; Ma, S.; He, J.; Gai, X.; Zhang, X. Ignition and Combustion Characteristic of B·Mg Alloy Powders. *Materials* **2022**, *15*, 2717. https://doi.org/10.3390/ma15082717

Academic Editor: Di Wu

Received: 28 February 2022
Accepted: 3 April 2022
Published: 7 April 2022

Publisher's Note: MDPI stays neutral with regard to jurisdictional claims in published maps and institutional affiliations.

Copyright: © 2022 by the authors. Licensee MDPI, Basel, Switzerland. This article is an open access article distributed under the terms and conditions of the Creative Commons Attribution (CC BY) license (https:// creativecommons.org/licenses/by/ 4.0/).

1. Introduction

As an energetic material, combustible metal powder is widely used in gunpowder, propellant, explosives, and pyrotechnics [1–6]. The reaction of metal powder and oxidant can provide the power for gun projectile launching and rocket missile propulsion. It is the power energy of the warhead explosion and the energy source that realizes smoke interference and combustion damage. The performance of metal powder to a large extent determines the level of combat technology of weapons and equipment, and is the core technology to achieve long-range strike and high-efficiency damage, and is the key material basis for the development of advanced weapons and equipment.

Among the combustible metal powders, magnesium is easy to ignite and burn but has a low calorific value. Compared with magnesium, boron has a very high calorific value. Its mass calorific value is 2.3 times that of magnesium and 1.9 times that of aluminum, and its volume calorific value is 3.09 times that of magnesium and 1.66 times that of aluminum. However, the ignition temperature of boron is high. In the combustion process, the liquid boron oxide that is generated hinders the reaction between boron and the oxidizer; moreover, the boron particles are easy to agglomerate, which makes the combustion difficult to continue. These factors lead to the low combustion efficiency of boron and underutilization of boron's high energy [3,7,8]. To solve the above problems, we prepared a new alloy material based on the synergistic effect with lower ignition temperature, faster combustion speed, weaker combustion agglomeration, and more complete combustion than a single metal powder material [9–12], important for improving the performance of propellants, explosives, and pyrotechnics.

The preparation of boron-containing alloys, such as Al–Ti–B and Ti–B alloys, has been reported by previous studies [13–17]. Shtessel et al. [10] prepared a series of Al–Mg, Al–Mg–H, and Al–B alloys by high-energy ball milling. Birol et al. [18] also prepared Al–B alloys from Al and B_2O_3 by ball milling, whereas Korchagin et al. [19] synthesized Ni_3B using high-energy ball milling of a mixture of Ni and B powders. Guo et al. prepared MgB_2 by a high-temperature sintering method at 1173 K [16,17].

Many methods based on differential thermal analysis (DTA) have been reported for estimating the ignition temperature of materials. In the mature and stable method, the ignition temperature is obtained by extrapolating the reaction temperature. However, this approach is subject to atmospheric factors, heating rate, and other parameters during testing. Shoshin et al. [20] studied the ignition of Al–Ti alloy in the air by coating a wire with the powder and measuring its temperature during electrical heating. Other researchers used lasers of different powers to measure the ignition and combustion characteristics [21–23]. Zhao [24] used a contact thermocouple to assess the ignition temperature of metal fuel in different atmospheres. Yang and co-workers [25] used planar flame burners to ignite modified boron and examine its combustion characteristics. Whittaker et al. [13] found that AlB_2 is an energetic fuel with a high heat of combustion. Arkhipov et al. [26] studied the ignition and combustion of propellants of Al, B, and Al–B powders with binders with oxidizing agents (ammonium perchlorate or ammonium nitrate).

In this study, MgB_2 was prepared by a combination of mechanical alloying and heat treatment, and the ignition and combustion of boron–magnesium alloys were studied with the ignition wire method and laser ignition infrared temperature measurement.

2. Materials and Methods

2.1. Materials and Instruments

Atomized spherical magnesium powder (Mg) has a particle size <45 μm and purity of 99.9%, Shanghai Shuitian Material Technology Co., Ltd., Shanghai, China. The amorphous boron powder (B) had a particle size of 1~3 μm and purity of 99.7%, Liaoning Yingkou Fine Chemical Company. MgB_2 was prepared by a combination of mechanical alloying and heat treatment.

The following instruments were used: a vibrating high-energy ball mill (Beijing Nonferrous Metals Research Institute, Beijing, China), a vacuum operation box (ZKX-2, Nanjing Nanda Instrument Factory, Nanjing, China), a tube furnace (SGL-1700-II, Shanghai Jujing Precision Instrument Manufacturing Co., Ltd., Shanghai, China), an X-ray diffractometer (D8 ADVANCE, Bruker, Karlsruhe, Germany), an S-4800 cold field emission scanning electron microscope (Hitachi, Tokyo, Japan), and a microcomputer automatic calorimeter (TRHW-7000E, Hebi City Tianrun Electronic Technology Co., Ltd., Hebi, China).

2.2. Preparation of MgB_2

The Mg and B powders in a mole ratio of 1:2 were placed in a stainless steel ball mill jar. The ball mill tank was sealed with an O-ring and passed through a flow of 0.1 MPa argon three times for three minutes. The vibrating high-energy ball mill used stainless steel balls of either φ10 mm or mixed sizes (φ2 mm, φ5 mm, and φ10 mm in a mass ratio of 1:1:3) at 20:1 mass to the powder materials. Powder loading and sampling operation were both carried out in a vacuum operation box filled with argon gas. During milling, the ball mill had a three-dimensional motion of rotational oscillation and vibration. The ball milling time was 12 min, and after milling, the powder was sealed in a vacuumed quartz glass tube and placed in a tube furnace with a flowing Ar atmosphere at 20 mL/min. The furnace was heated from room temperature (298 K) to 853 K at a rate of 5 K/min and held for 10 h.

2.3. Performance Test

The structure of the product was analyzed by X-ray diffraction (XRD) using Cu target Kα radiation (0.15406 nm), a working voltage of 40 kV, a working current of 20 mA, and 2θ angle scan range of 10–90° in steps of 0.02° at a scanning speed of 0.5°/min.

The setup for measuring the ignition temperature is shown in Figure 1. The sample powder was placed on the wound ignition wire. After the power was turned on, the ignition wire was gradually heated, and we assume that at the point of powder ignition the powder and the wire had the same temperature. The timing of the ignition was monitored by a photodetector, and the temperature change in the ignition wire was measured by an infrared thermocouple. To avoid the heat released by sample combustion interfering with the temperature measurement, the infrared thermocouple monitored other parts of the ignition wire not covered with the sample powder. The light intensity and temperature data during the experiments were recorded by the data acquisition instrument.

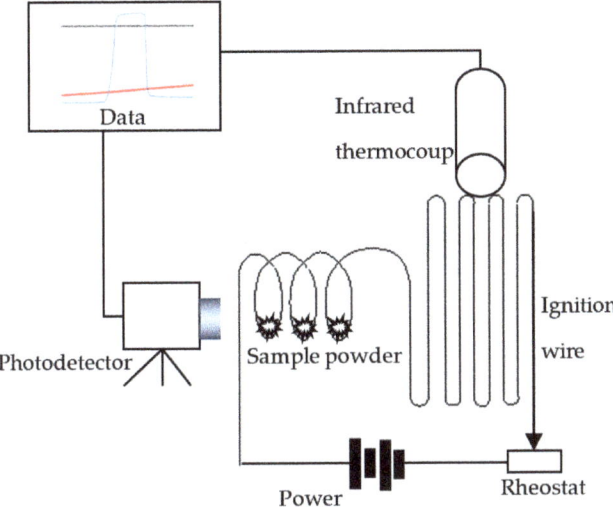

Figure 1. The sketch of the ignition temperature tester.

The ignition wire was a 0Cr27Al7Mo2 iron chrome aluminum alloy wire 0.5 mm in diameter. An SYS480S36 DC switching power supply with a maximum output of 480 W directly heated the ignition wire. The supply had overcurrent protection, and its current was not less than 10 A to provide instantaneous large current through the ignition wire. To facilitate the determination of the ignition starting time, the DC power supply was controlled by a solid-state relay (SSR200D40, Hangzhou Guojing Technology, Hangzhou, China). The temperature was measured using an infrared thermocouple (OS37-20-K, OMEGA Company, Austin, TX, USA) supplemented by a temperature transmitter (TXDIN1620, OMEGA Company, Austin, TX, USA), control line, and switching power supply (YMGUD-2030LIXA, Yongming Power Company, Taiyuan, China). The infrared thermocouple output was the K-type, the temperature range was 533–1923 K, and the temperature accuracy was ±2% in the linear zone and ±5% in the other ranges. The spectra of combustion for Mg, B, and MgB_2 powders had a wavelength range of 300–900, 450–650, and 300–800, respectively, with peak wavelengths of 497.7, 546, and 546.8/579 nm. The test equipment used Thorlabs' DET10A photodiode, which is a silicon-based photodetector with a detection wavelength covering the visible range and a peak response at 730 nm in the near-infrared region. The data acquisition was carried out with a NI9222 card (National Instruments, Austin, TX, USA) with a maximum sampling rate of 500 K, and LABVIEW software.

The laser ignition system consisted of three parts: laser igniter, combustion chamber, and data acquisition device, as shown in Figure 2. The laser igniter included three main parts: fiber-coupled semiconductor laser, laser power supply, and an optical coupler. The microscopic high-speed photography-infrared thermography synchronization device recorded the flame, the sample morphology, and the temperature field during the MgB_2 ignition and combustion processes. The thermal imaging camera (SC325, FLIR Systems, Inc., Tigard, OR, USA) could record in a temperature range up to 2273 K. The high-speed camera (FASTCAM-APX-RS, Photron, Arizona, CA, USA) had an AF Micro-Nikkor 60 mm f/2.8 D macro lens. To facilitate the observation of the ignition and combustion process of small particles, an AF Micro-Nikkor 60 mm f/2.8 D lens with macro function (Nikon Corporation, Tokyo, Japan) was used. In the experiment, a continuous laser with a power of 4 W was used to ignite and burn 50 mg of micro-clustered sample in an air atmosphere of 0.1 MPa. Infrared thermal imagers and high-speed cameras simultaneously detected changes in the flame topography and temperature field. The surface morphology of the particles was directly observed by high-speed photography plus the macro lens, and the shooting speed was 2000 frames per second.

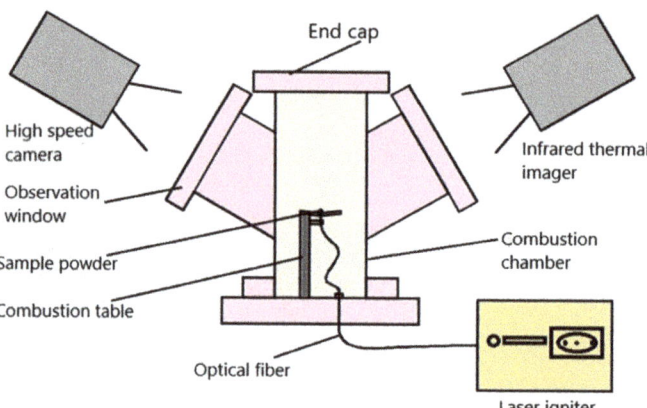

Figure 2. Schematic diagram of laser ignition experimental system.

The micro-machine automatic calorimeter was used in an oxygen atmosphere of 3 MPa. The volume of the oxygen bomb was 300 mL, and the sample amount of the single test was 0.5 g. The combustion heat tests were performed on the samples of Mg, B, and MgB_2.

3. Results

3.1. Preparation and Characterization of MgB_2

The XRD spectrum of the prepared MgB_2 was compared with the PDF card standard spectrum by XRD spectrum software Jade, as shown in Figure 3.

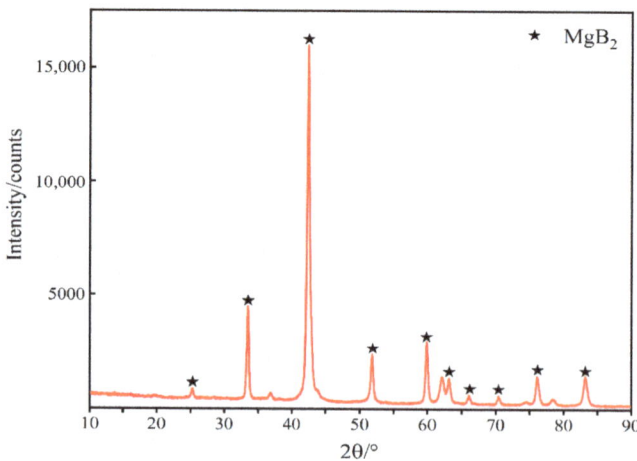

Figure 3. Comparison of the XRD patterns of prepared MgB_2 and PDF Card.

The position of the spectrum peak and the corresponding intensity are consistent with the standard spectrum peak in the PDF card, and there is basically no spurious peak, indicating that the product is MgB$_2$ with high purity.

3.2. Experimental Study of MgB$_2$ Ignition Performance

In this paper, the test device was designed to completely close the switch, and the resistance wire started to heat up as the starting moment; that is, the solid-state relay was fully turned on, and the control voltage continued to be higher than 3.5 V as the heating start time. Theoretically, the temperature and intensity of the powder change during the powder ignition process, and the temperature at the moment of the mutation is the ignition temperature. Since the temperature of the heating wire is increased after the electric wire is energized, it gradually becomes brighter. It is known that the output voltage of the photodiode is not more than 0.3 V during the heating process of the ignition wire. Therefore, the design of the test device is such that the light intensity continuously exceeds 0.3 V as the ignition starting time. The first peak of the light intensity is the moment when the powder is completely ignited, and the average temperature of the corresponding ignition wire is the powder ignition temperature. Since the infrared thermocouple test delay is 80 ms, the temperature curve is shifted back from the start time by 80 ms during data processing. The period of temperature change of the ignition wire is about 20 ms; that is, the infrared thermocouple outputs effective temperature measurement data after continuous testing for 20 ms. If the powder ignition time is less than 20 ms after ignition, start time is taken as the ignition duration, and the corresponding ignition wire temperature is the ignition temperature. The obtained magnesium powder ignition temperature curve is shown in Figure 4.

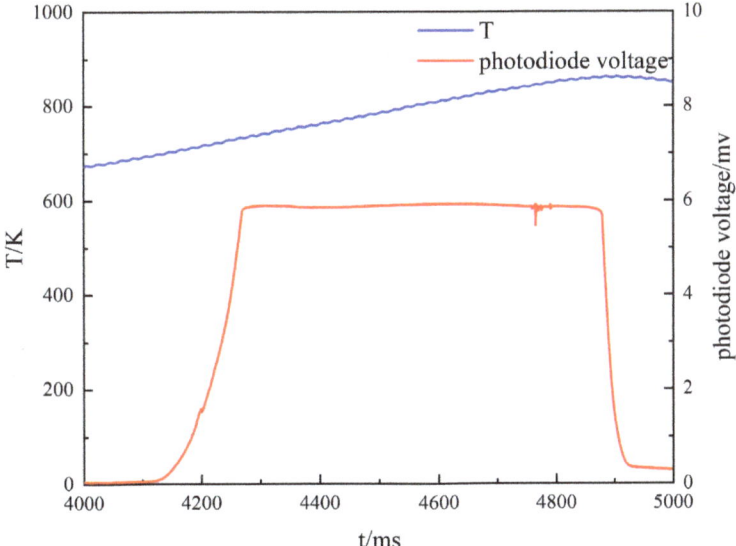

Figure 4. The ignition temperature curves of Mg powder.

The blue curve is the temperature, and the red curve is the voltage value of the photodiode. It can be seen from the test that the average ignition temperature of magnesium powder is 719 K, which is the same value reported in the literature [25]. Therefore, the feasibility of the test device is verified by the ignition temperature test of magnesium powder.

The procedure of the MgB$_2$ ignition temperature test included adjusting the varistor shown in Figure 1 and reducing the loop load resistance by 1 Ω. The obtained curve shows that the ignition temperature of the boron-magnesium alloy is 1292 K, as shown in Figure 5.

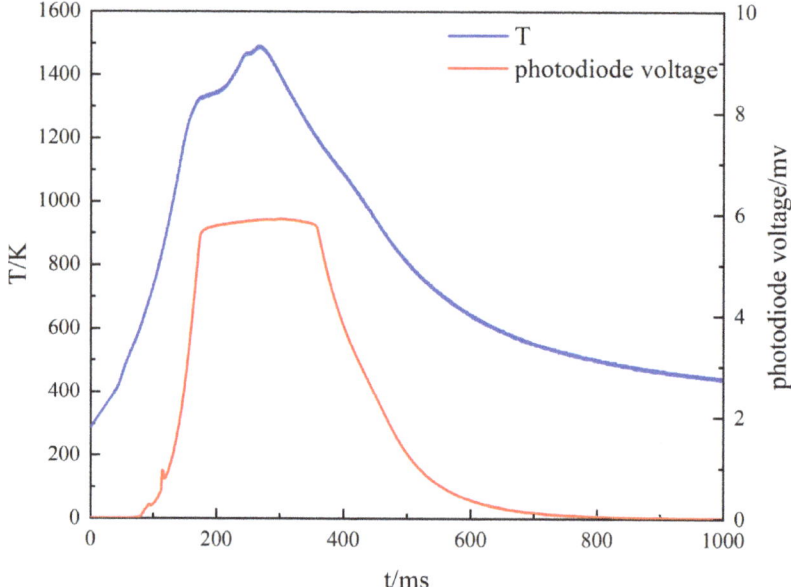

Figure 5. Curves of MgB$_2$ ignition temperature test.

In the laser ignition experimental system shown in Figure 2, the micro-cluster of the MgB$_2$ sample was placed on the combustion table. The laser fiber was adjusted to face the center of the micro-cluster, and then the focal length of the macro lens was adjusted to ensure that the micro-cluster was in the focus position, while simultaneously turning on the infrared thermal imager. The ignition experiment started by triggering the high-speed camera and turning on the laser power. The temperature during ignition and combustion of the MgB$_2$ sample in the infrared thermal imaging test is shown in Figure 6 and the infrared temperature measurements at different time points are shown in Table 1 and Figure 7.

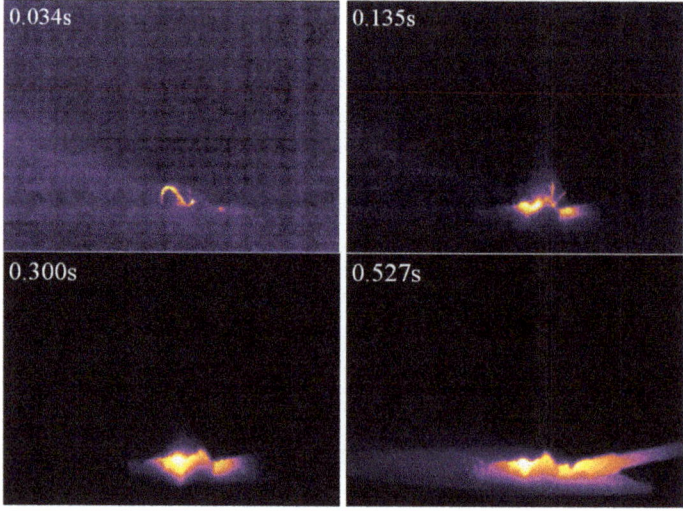

Figure 6. Infrared temperature measurements during ignition and combustion of the of the MgB$_2$ sample.

Table 1. Infrared temperature of the MgB$_2$ sample at different time points.

t(s)	0	0.034	0.068	0.135	0.174	0.209	0.300	0.333	0.527
T(K)	290	381	445	617	705	730	969	1293	1163

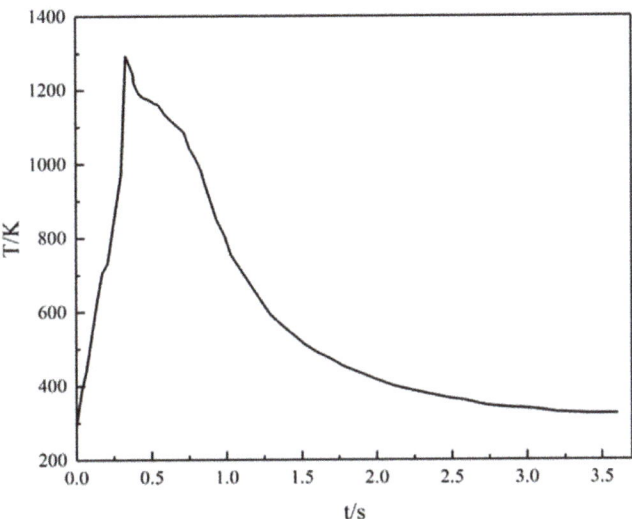

Figure 7. The curve of MgB$_2$ sample ignition temperature.

The MgB$_2$ sample reached the highest ignition temperature of 1293 K at 0.333 s, and a flame appeared on the surface of the sample. Macek et al. studied the ignition and combustion of boron particles which showed that the ignition temperature of boron is about 2000 K under the air atmosphere [27]. The boron powder is surrounded by the highly viscous boron oxide B$_2$O$_3$ produced during the ignition and combustion process, so that the ignition and combustion of boron powders are difficult to carry out. Compared with boron, the MgB$_2$ sample ignition temperature was lower by 700 K, indicating that the ignition performance is significantly improved after boron alloying.

3.3. Study of the Combustion Properties of MgB$_2$

Infrared thermal imagers and high-speed cameras were used to simultaneously detect changes in flame topography and temperature field distribution. Among them, the initial ignition timing of the sample is t_0, and the continuous burning time of the sample is t. A comparison of the high-speed photographic test results of the ignition reactions of Mg, B, and MgB$_2$ samples is shown in Figure 8.

It can be seen from Figure 8 that the flame size of Mg is large, and the flame size of B is small during the combustion process, and the flame morphology does not change much during the combustion process, whereas the flame of the MgB$_2$ sample is large at the initial stage. The flame of the MgB$_2$ sample becomes smaller after 2/3 t, but the duration is still longer. This is because Mg has a low melting point and a low boiling point, and is easily melted and vaporized during ignition and combustion, so that gaseous Mg is easily formed during combustion. Boron has a relatively high melting point and boiling point, and is not easily vaporized during ignition and combustion, and its combustion process is close to solid phase combustion. The combustion of the MgB$_2$ sample can be seen as a two-stage process. During the ignition and initial combustion stages, the MgB$_2$ sample can form part of the gaseous Mg, thus forming a larger flame, and then approaching the solid phase combustion with B.

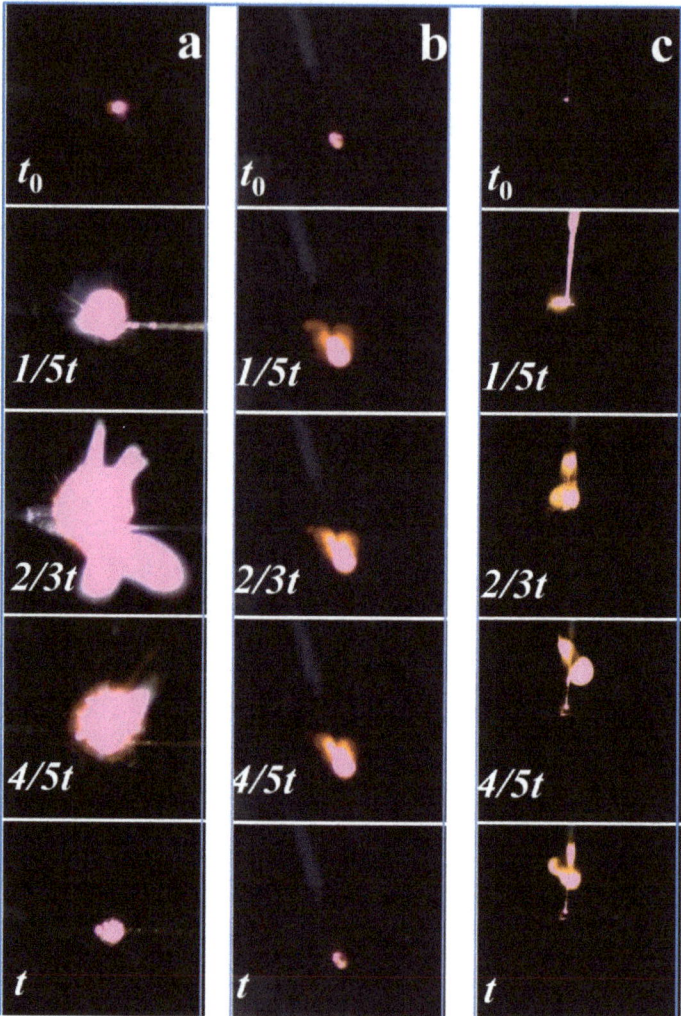

Figure 8. Burning flame morphology of Mg (**a**), B (**b**), and MgB$_2$ (**c**) samples.

The ignition experiments of the Mg, B, and MgB$_2$ samples were repeated five times, combined with the infrared test data, and the characteristic parameters thereof are shown in Table 2. Mg burns fastest whereas B burns the slowest. MgB$_2$ sample burns at a medium speed, but its maximum combustion temperature is close to Mg.

Table 2. Characteristic ignition parameters of Mg, B, and MgB$_2$ samples.

Parameter Sample	Mg	B	MgB$_2$
Burning time(s)	0.41	0.86	0.53
Ignition temperature(K)	719	2000	1293

4. Conclusions

An ignition study of the boron–magnesium alloy was carried out by using the ignition wire method and the laser ignition infrared temperature measurement method. Compared with boron, the ignition temperature of MgB$_2$ is greatly reduced, suggesting improved ignition performance. The main conclusions obtained can be summarized as:

1. The results show the maximum ignition temperatures of MgB$_2$ obtained by the two methods are 1292 K and 1293 K, respectively.
2. The calculation results of temperature change during MgB$_2$ laser ignition were in good agreement with the experimental results, indicating that the calculation model is capable of describing the temperature change of the particle ignition process.
3. The flame morphology and combustion state are close to the combustion with gaseous Mg in the initial stage. However, the subsequent combustion process is close to the combustion process of B. Compared with Mg and B, the MgB$_2$ sample burns at a medium speed.

Author Contributions: Conceptualization, Y.M., K.Z. and X.Z.; methodology, Y.M., K.Z. and X.Z.; investigation, S.M., J.H., X.Z. and Y.M.; writing—original draft preparation, Y.M. and K.Z.; writing—review and editing, X.Z.; supervision, X.G. and S.M. All authors have read and agreed to the published version of the manuscript.

Funding: This research was funded by the National Natural Science Foundation of China (Grant No. 51404279).

Institutional Review Board Statement: Not applicable.

Informed Consent Statement: Not applicable.

Data Availability Statement: The data presented in this study is available on request from the corresponding author.

Conflicts of Interest: The authors declare no conflict of interest.

References

1. Beckstead, M.W. Correlating aluminum burning times. *Combust. Explos. Shock Waves* **2005**, *41*, 533–546. [CrossRef]
2. Berner, M.K.; Zarko, V.E.; Talawar, M.B. Nanoparticles of energetic materials: Synthesis and properties. *Combust. Explos. Shock Waves* **2013**, *49*, 625–647. [CrossRef]
3. Pang, W.; Xue, Z.F.; Zhao, F.Q. *Boron-Based Fuel Rich Solid Rocket Propellant Technolog*; National Defense Industry Press: Beijing, China, 2016; pp. 19–21.
4. Kim, S.; Lim, J.; Lee, S.; Jeong, J.; Yoon, W. Study on the ignition mechanism of Ni-coated aluminum particles in air. *Combust. Flame* **2018**, *198*, 24–39. [CrossRef]
5. Zhu, B.; Li, F.; Sun, Y.; Wu, Y.; Shi, W.; Han, W.; Wang, Q.; Wang, Q. Enhancing ignition and combustion characteristics of micron-sized aluminum powder in steam by adding sodium fluoride. *Combust. Flame* **2019**, *205*, 68–79. [CrossRef]
6. Feng, Y.; Xia, Z.; Huang, L.; Ma, L. Ignition and combustion of a single aluminum particle in hot gas flow. *Combust. Flame* **2018**, *196*, 35–44. [CrossRef]
7. Glassman, I.; Williams, F.A.; Antaki, P. A physical and chemical interpretation of boron particles combustion. In Proceedings of the 20th Symposium (International) on Combustion, Pittsburgh, PA, USA, 12–17 August 1984.
8. Chen, B.H.; Liu, J.Z.; Liang, D.L.; Zhou, Y.N.; Zhou, J.H. Research progress in coating mechanism and technology of boron particles. *Chin. J. Explos. Propellants* **2016**, *39*, 13–21.
9. Shoshin, Y.L.; Mudryy, R.S.; Dreizin, E.L. Preparation and characterization of energetic Al-Mg mechanical alloy powders. *Combust. Flame* **2002**, *128*, 259–269. [CrossRef]
10. Shtessel, E.; Dreizin, E. *High Energy Metallic Mechanical Alloys for New Explosives and Incendiary Devices with Controllable Explosion Parameters, ADA397097*; Exotherm Corporation: Camden, NJ, USA, 2001.
11. Aly, Y.; Schoenitz, M.; Dreizin, E.L. Ignition and combustion of mechanically alloyed Al-Mg powders with customized particle sizes. *Combust. Flame* **2013**, *160*, 835–842. [CrossRef]
12. Aly, Y.; Dreizin, E.L. Ignition and combustion of Al·Mg alloy powders prepared by different techniques. *Combust. Flame* **2015**, *162*, 1440–1447. [CrossRef]
13. Whittaker, M.L.; Cutler, R.A. Effect of synthesis atmosphere, wetting, and compaction on the purity of AlB2. *J. Solid State Chem.* **2013**, *207*, 163–169. [CrossRef]
14. Jia, Y.J. Status of Production Technology and Function mechanism of Al-Ti-B Alloy Products. *Nonferrous Met. Process.* **2017**, *46*, 33–34.
15. Jalaly, M.; Gotor, F.J. A new combustion route for synthesis of TaB2 nanoparticles. *Ceram. Int.* **2018**, *44*, 1142–1146. [CrossRef]
16. Guo, Y.; Zhang, W.; Zhou, X.; Bao, T. Magnesium Boride Sintered as High-energy Fuel. *J. Therm. Anal.* **2013**, *113*, 787–791. [CrossRef]

17. Guo, Y. *Study on Preparation, Combustion Property and Application of Modified Boron*; National University of Defense Technology: Changsha, China, 2014.
18. Birol, Y. Aluminothermic reduction of boron oxide for the manufacture of Al-B alloys. *Mater. Chem. Phys.* **2012**, *136*, 963–966. [CrossRef]
19. Korchagin, M.A.; Dudina, D.V.; Bokhonov, B.B.; Bulina, N.V.; Ukhina, A.V.; Batraev, I.S. Synthesis of nickel boride by thermal explosion in ballmilled powder mixtures. *J. Mater. Sci.* **2018**, *53*, 13592–13599. [CrossRef]
20. Shoshin, Y.L.; Trunov, M.A.; Zhu, X.; Schoenitz, M.; Dreizin, E.L. Ignition of aluminum-rich Al-Ti mechanical alloys in air. *Combust. Flame* **2006**, *144*, 688–697. [CrossRef]
21. Badiola, C.; Gill, R.J.; Dreizin, E.L. Combustion characteristics of micron-sized aluminum particles in oxygenated environments. *Combust. Flame* **2011**, *158*, 2064–2070. [CrossRef]
22. Gill, R.J.; Badiola, C.; Dreizin, E.L. Combustion times and emission profiles of micron-sized aluminum particles burning in different environments. *Combust. Flame* **2010**, *157*, 2015–2023. [CrossRef]
23. Yu, B.; Du, C.Z. Study on the Temperature Determination Technique for Ignition and Combustion of Propellants. *Chin. J. Explos. Propellants* **2001**, *24*, 28–31.
24. Qin, Z.; Paravan, C.; Colomobo, G.; Deluca, L.T.; Shen, R.Q.; Ye, Y.H. Ignition Temperature of Metal Fuel in Different Atmosphere. *Initiat. Pyrotech.* **2014**, *4*, 24–27.
25. Yuan, C.M.; Li, C.; Li, G.; Zhang, P.H. Ignition temperature of magnesium powder clouds: A theoretical model. *J. Hazard. Mater.* **2012**, *239–240*, 294–301.
26. Arkhipov, V.A.; Zhukov, A.S.; Kuznetsov, V.T.; Zolotorev, N.N.; Osipova, N.A.; Perfil'eva, K.G. Ignition and combustion of condensed systems with energy fillers. *Combust. Explos. Shock Waves* **2018**, *54*, 689–697. [CrossRef]
27. Maček, A. Combustion of boron particles: Experiment and theory. In Proceedings of the 14th Symposium (International) on Combustion, University Park, PA, USA, 20–25 August 1972.

Review

Review of the Effect of Surface Coating Modification on Magnesium Alloy Biocompatibility

Xuan Guo, Yunpeng Hu, Kezhen Yuan and Yang Qiao *

School of Mechanical Engineering, University of Jinan, Jinan 250022, China; guoxuan123125@163.com (X.G.); hhyypp82516@163.com (Y.H.); y2294_6050@163.com (K.Y.)
* Correspondence: me_qiaoy@ujn.edu.cn; Tel.: +86-152-7510-6865

Abstract: Magnesium alloy, as an absorbable and implantable biomaterial, has been greatly developed in the application field of biomaterials in recent years due to its excellent biocompatibility and biomechanics. However, due to the poor corrosion resistance of magnesium alloy in the physiological environment, the degradation rate will be unbalanced, which seriously affects the clinical use. There are two main ways to improve the corrosion resistance of magnesium alloy: one is by adding alloying elements, the other is by surface modification technology. Compared with adding alloy elements, the surface coating modification has the following advantages: (1) The surface coating modification is carried out without changing the matrix elements of magnesium alloy, avoiding the introduction of other elements; (2) The corrosion resistance of magnesium alloy can be improved by relatively simple physical, chemical, or electrochemical improvement. From the perspective of corrosion resistance and biocompatibility of biomedical magnesium alloy materials, this paper summarizes the application and characteristics of six different surface coating modifications in the biomedical magnesium alloy field, including chemical conversion method, micro-arc oxidation method, sol-gel method, electrophoretic deposition, hydrothermal method, and thermal spraying method. In the last section, it looks forward to the development prospect of surface coating modification and points out that preparing modified coatings on the implant surface combined with various modification post-treatment technologies is the main direction to improve biocompatibility and realize clinical functionalization.

Keywords: magnesium alloy; biocompatibility; surface coating modification; corrosion resistance; implantable bio-metal materials

1. Introduction

Medical metal materials play an important role in the repair and replacement of bone defects caused by bone joint diseases and trauma [1]. Magnesium is a macroelement needed in the human body, so the degradation of magnesium in the human body is generally considered as non-toxic [2]. At the same time, magnesium alloy can be dynamically degraded after being implanted into the human body, so excessive Mg^{2+} will be discharged out of the human body through the circulation system [3]. In addition, the elastic modulus of magnesium alloy implanted in the human body is similar to that of bone, so the stress shielding effect can be improved to a certain extent [4,5]. Compared with polymers, ceramics, and other medical implantable metals, magnesium alloy materials have good human biochemical compatibility, unique degradability, and reliable mechanical properties [6,7]. However, because the chemical properties of medical magnesium alloy are too active and the corrosion rate in the human body is too fast, the supporting strength of the material decreases sharply in a short time, and the biocompatibility is very poor, which affects its application effect [8]. Moreover, as a heterogeneous material, its safety needs to be further improved before it can be widely used in clinical orthopedics [9]. At present, research shows that the surface coating modification of magnesium alloy can not only improve the corrosion resistance of magnesium alloy but also improve its mechanical properties and biocompatibility, so it is of great significance to master the surface coating modification of different magnesium alloys [10–12]. In recent years, with the progress of research, the surface structure has been considered as one of the important factors affecting the combination of implants and bones [13,14], and its morphology can regulate the growth orientation of cells [15], and processing controllable surface structure is of great significance to improving the biocompatibility of materials [16]. Based on previous studies, this paper discusses the characteristics of different surface coating modification technologies of magnesium alloys and their effects on corrosion resistance and biocompatibility of magnesium alloys and discusses the future development trend of surface coating modification of biomedical magnesium alloys in the future.

2. Effect of Surface Coating Modification on Biocompatibility of Magnesium Alloy

Surface modification technology refers to the formation of protective coatings with different functions on the surface of substrate materials by various technological methods to achieve the desired purpose, which is a very important surface modification method in surface engineering technology [17]. A barrier can be formed by adding coating treatment on the metal surface to control the degradation rate of magnesium alloy [18]. The biocompatibility of magnesium alloy mainly means that magnesium alloy, as a bone implant, will not bring adverse reactions to the body on the basis of realizing connection, fixation, and maintaining normal healing of the body. In the whole healing process, magnesium alloy must first ensure that it can support the fracture site and ensure the normal healing of bones [19]. Second, the process of gradual corrosion and degradation of magnesium alloy in the human body should be at a moderate speed. Too fast will cause the bone not to heal in a short time, which will lead to treatment failure. Too slow will cause the patient's treatment time to be too long [20], and the substances produced by corrosion and degradation should not be harmful to the human body. For example, hydrogen generated by the chemical reaction between magnesium electrochemical potential and electrolytes in an aqueous environment will lead to inflammation [21]. Finally, on the premise of ensuring basic function, magnesium alloy is treated on its surface to make it beneficial to cell growth, thus further improving biocompatibility. Common metal surface coating treatment methods include chemical conversion method, sol-gel method, micro-arc oxidation method, electrodeposition method, and hydrothermal treatment.

3. Effect of Chemical Transformation on Biocompatibility

Chemical conversion method is a method of converting the surface of the material itself into a coating, which belongs to the in-situ growth of the material and has good coating adhesion [22]. In the process of conversion coating, the substrate to be protected is immersed in the solution that reacts with the surface, changing the concentration of metal ions and the pH value of the interface of the metal solution so that the substrate material itself is converted into another material, forming a coating with good adhesion [23]. Chemical conversion method is a method with convenient production, small upfront investment, quick response, and controllable production conditions. Because the surface of the material itself is converted into a coating, the ceramic film has a strong bonding force, and it can transform one nano-material into another new nano-structure that is difficult to prepare directly, complex, and has unique properties. Therefore, this technology has a broad application prospect. Chemical conversion coating usually includes chromate coating, phosphate coating, fluorine-containing coating, etc. However, considering that chromate will leach out of the coating and lead to cancer, the coatings of chemical conversion method mainly focus on phosphate coating and fluorine-containing coating.

3.1. Study on Phosphate Coating

In the study of biocompatibility of phosphate coating, the first thing to study is cell culture in vitro. Xu et al. [24] coated calcium phosphate coating on magnesium alloy surfaces by the chemical transformation method, and cell line L929 showed good growth rate and proliferation characteristics in in vitro cell experiments. Figure 1 shows the contrast diagram of new bone cells that were not coated with calcium phosphate coating after 4 weeks of culture. Calcium phosphate coated magnesium alloy was implanted in vivo to study the early bone reaction. The results showed that calcium phosphate coated magnesium alloy showed significantly improved bone conduction and osteogenesis ability in the first four weeks after operation. Calcium phosphate coated magnesium alloy provided significantly good surface bioactivity for magnesium alloy and promoted the early bone growth at the implant/bone interface. Lorenz et al. [25] studied the effect of calcium/magnesium phosphate coating on the surface of magnesium on the survival rate of mouse fibroblasts in cell culture experiments. After soaking in NaOH-SBF, a mixed calcium/magnesium phosphate coating was formed on pure magnesium. Compared with the magnesium surface simply soaked in NaOH solution, the cell survival rate was improved. Figure 2 shows the change of cell fluorescence imaging and cell diffusion area with culture time. The results show that chemical transformation with NaOH-SBF solution can improve the reactivity, chemical properties, and roughness of the material surface and provide a feasible strategy for cell survival on the magnesium surface. With the in-depth study of biocompatibility of magnesium alloy, people began to introduce implants into animals for research. Yang et al. [26] coated calcium phosphate coating on magnesium alloy (AZ31). The samples were implanted into rabbits to study the early bone reaction. This is because the slow biodegradation rate confirms the positive effect of phosphate coating.

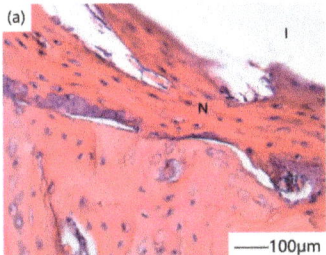
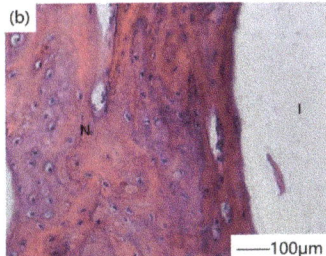

Figure 1. Comparison diagram of newly-born bone cells coated with (**a**,**b**) calcium phosphate around the culture. Reprinted with permission from Ref. [24]. Copyright 2019 Copyright Elsevier.

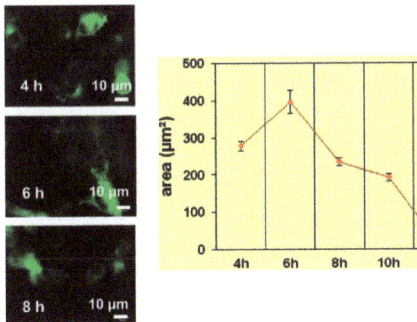

Figure 2. Mouse fibroblasts on Mg surface that had been pre-treated by soaking in M-SBF at 37 °C for 5 days. Fluorescence imaging of cells and cell spreading area as a function of culture time. Reprinted with permission from Ref. [25]. Copyright 2009 Copyright Elsevier.

To sum up, the reason why phosphate coating has good biocompatibility is that after the implant is implanted into the animal body, the surface begins to dissolve in body fluid and releases a large amount of calcium and phosphorus ions, which partially supersaturates the body fluid ions near the bone, which is beneficial to the formation and growth of bone cells. Furthermore, with the increase in time, the concentration of Mg^{2+} in body fluid increases, and it combines with calcium ions to form the nucleation of Ca-P compounds, which leads to the growth of Ca-P on magnesium alloy, which is beneficial for osteoblasts to adhere to the phosphate surface. Obviously, phosphate conversion coating has high temperature resistance, chemical stability, and excellent biocompatibility and is a good substitute for chromate as a biological coating. However, the phosphate conversion coating prepared on the surface of magnesium alloy by chemical conversion method generally has defects such as porosity and cracks, and the solution pollutes the environment, resulting in a high cost of treating the three wastes, which does not meet the needs of high-quality clean production and sustainable development strategy in China.

3.2. Study on Fluorine-Containing Coating

In the research on fluorine-containing coating, the mechanism that magnesium fluoride coating can delay corrosion is mainly used. In the research on cell culture of magnesium alloy in vitro, Carboneras et al. [27] studied the biodegradation kinetics of powder metallurgy magnesium, cast magnesium, and magnesium alloy AZ31, which was evaluated by electrochemical impedance spectroscopy measurement in cell culture medium (DMEM). In order to reduce their degradation rate, chemical conversion treatment was carried out in hydrogen fluoride during the experiment to form magnesium fluoride coating. Figure 3 shows the SEM surface morphology of magnesium fluoride coating material after immersion in DMEM for 11 days. The results showed that magnesium fluoride coating slowed down the biodegradation rate, especially on cast magnesium and magnesium alloy AZ31, and delayed the corrosion time for at least one week. This is because the formed coating has inherent protective properties, such as high density, low water solubility, and high insulation caused by high impedance. Chiu et al. [28] used pure magnesium as the substrate and soaked it in hydrofluoric acid at room temperature to obtain a dense and crack-free magnesium fluoride coating with a thickness of 1.5μm on the surface. Magnesium fluoride coating is mainly composed of tetragonal magnesium difluoride, and the crystallite size is estimated to be several nanometers. The results show that with a good electrode impedance of 0.18 kΩcm^2, the corrosion current density is reduced by 40 times. This is because the magnesium fluoride coating is chemically inert and can be used as an anticorrosive barrier coating. In a word, chemical conversion treatment of fluorinated coating is a simple and promising method to improve the corrosion resistance of magnesium in Hank's solution, and it can also be used as the pretreatment step of subsequent coating. In the exploration of magnesium alloy as an implant in animals, Witte et al. [29] studied whether the extruded magnesium alloy LAE442 coated with magnesium fluoride was corroded and degraded by appropriate host reaction in rabbits. The results show that the fluorine-coated implant can effectively delay the corrosion of magnesium alloy and has an acceptable host reaction in rabbits. This is because the dense magnesium fluoride coating plays an isolation role, effectively reducing the hydrogen generated by the reaction between body fluid and magnesium alloy and reducing the toxic effect caused by hydrogen evolution. Moreover, the corrosion products formed in the coating materials are compounds rich in calcium and phosphorus, which are necessary for bone health, which also confirms the clinical application of magnesium fluoride coating as biodegradable implant. However, magnesium fluoride (MgF_2) coating seems to stimulate local synovial tissue during the dissolution process, and pitting corrosion is larger than that of uncoated alloy after the coating is broken.

Fluoride conversion coating has good corrosion resistance, improves cell response and biocompatibility, and is an effective coating widely used in the biomedical field. However, although fluoride conversion coating can provide protection for magnesium alloy at the initial stage of implantation, it is difficult to provide long-term effective corrosion protection for magnesium alloy because the coating is very thin. At the same time, because excessive fluoride will have a negative impact on human bones, and the release process of fluoride ions during the degradation of magnesium implants and its toxicity to surrounding tissues are still unclear, the safety of fluoride coating still needs many clinical experiments to study.

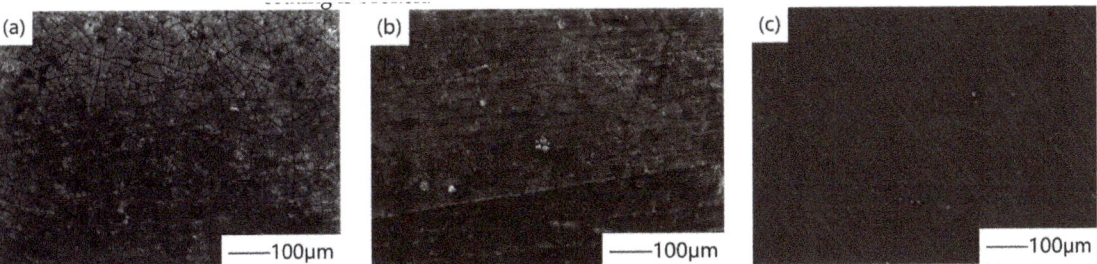

Figure 3. SEM surface morphology of the magnesium fluoride-coated materials after 11 days of immersion in DMEM: (**a**) PM Mg, (**b**) cast Mg, and (**c**) AZ31 alloy. Reprinted with permission from Ref. [27]. Copyright 2011 Copyright Elsevier.

4. Effect of Micro-Arc Oxidation on Biocompatibility

Micro-arc oxidation (MAO), also known as plasma electrolytic oxidation (PEO) or micro-plasma oxidation (MPO), is a high-pressure plasma-assisted anodizing process [30]. Micro-arc oxidation has the following advantages: (1) Improved surface hardness, with microhardness of 1000 to 2000 HV, up to 3000 HV, which is comparable to that of cemented carbide; (2) good wear resistance, which fundamentally overcomes the shortcomings of soft metal materials such as aluminum and magnesium alloy in application; (3) good insulation performance, with insulation resistance up to 100 mΩ; (4) the solution is environmentally friendly and meets the requirements of environmental protection discharge. Micro-arc oxidation electrolyte does not contain toxic substances and heavy metal elements, and it has strong anti-pollution ability and a high recycling rate, thus causing little environmental pollution; (5) the reaction is carried out at room temperature, which is convenient to operate and easy to master. In the coating surface modification process, the micro-arc oxidation process is often used for titanium alloy, and it plays an important role. In recent years, both of them have developed rapidly. The similarities are an alkaline oxidation environment, the conventional oxidation system includes phosphate electrolyte, and there is little difference in arc voltage and oxidation speed [31,32]. Moreover, after oxidation at 350–360 V for 15–20 min, the corrosion current of MAO coating on titanium alloy surface is lower than that of magnesium alloy, and the corrosion voltage is higher than that of magnesium alloy, so titanium alloy coating is more corrosion resistant [33,34], which also includes the application of titanium alloy. However, magnesium alloy is more suitable for bone plate because of its unique degradability, which can degrade in vivo to avoid the pain of secondary removal.

The morphology and chemical properties of MAO coating depend not only on the composition of electrolyte and properties of alloy, but also on processing parameters such as current density [35], voltage [36], heat treatment time [37], power supply mode, and loading parameters [38]. Generally, MAO coating on magnesium alloy consists of a porous outer layer and a thin barrier inner layer. Due to the discharge caused by the current breaking through the local growth layer in the process of micro-arc oxidation, characteristic micropores are produced on the surface of the coating. This porous coating provides certain corrosion protection for the magnesium substrate, and it can also act as an intermediate layer to improve the bonding force of the composite coating [39].

Yu et al. [40] formed ceramic bioactive coating on ZK61 magnesium alloy substrate by micro-arc oxidation, and the static corrosion test by soaking in simulated body fluid (SBF) verified that the coating could improve the corrosion resistance of the substrate. However, the static corrosion medium studied by Yu cannot effectively and truly approach the dynamic physiological flow state of body fluid in the human body. Han et al. [41] designed the flow field to simulate the biological corrosion performance of the human body in real physiological environment, providing a new scientific theoretical basis. However, the characteristic micropores formed by micro-arc oxidation also provide a path for corrosive solution, which leads corrosive ions to contact the magnesium matrix and react with it. The solutions are usually divided into two categories, one is to form self-sealing coating by adding elements or particles, and the other is to carry out post-treatment by a two-step method combined with other processes. Ding et al. [42] prepared a ceramic coating containing hydroxyapatite (HA) on biodegradable $Mg_{66}Zn_{29}Ca_5$ magnesium alloy by micro-arc oxidation by adding hydroxyapatite (HA) particles in the electrolyte. Figure 4 shows the scanning electron microscope image of the micro-arc oxidation coating. The results showed that the biocompatibility was the best when HA concentration was 0.4 g/L. Li et al. [43] showed that by adding rare earth elements, the corrosion resistance is improved by reducing the active area of the substrate surface. With the deepening of research, in recent years, strontium [44], gallic acid [45], and copper [46], which can enhance bone formation and antibacterial effects, have been added to electrolyte solutions.

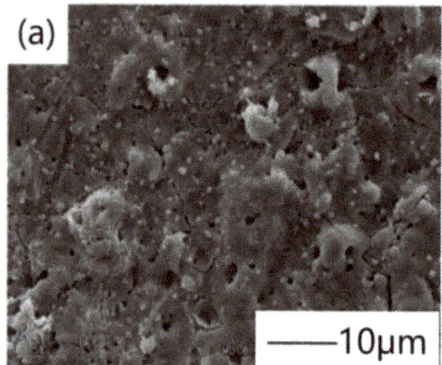 

Figure 4. SEM images of blood platelet adhesion on the micro-arc oxidation (MAO) coating HA-containing coating: (**a**) SEM images with scale of 10μm, (**b**) SEM images with scale of 2μm. Reprinted with permission from Ref. [42]. Copyright 2018 Copyright Elsevier.

Post-treatment is carried out by a two-step method combined with other processes, one of which is chemical treatment such as the sol-gel method and electrophoretic deposition method, which will be summarized below. The other is physical treatment, such as sliding friction technology and ultrasonic cold forging. Hu et al. [47] combined sliding friction technology (SFT) developed by our research group with MAO, Yang et al. [48] combined with MAO by ultrasonic cold forging (UCFT), and Lavanya et al. [49] combined it with powder coating double layer, all of which improved the corrosion resistance of magnesium alloys.

The reason why micro-arc oxidation coating can delay corrosion is that magnesium alloy with MAO coating is immersed in Hanks, with the increase in soaking time, the thin barrier inner layer will prevent further penetration of the corrosion medium, resulting in slow corrosion rate. Moreover, in Hanks, the corrosion products obtained by spontaneous mineralization can realize the self-protection of the coating and slow down the corrosion rate of the coating. By improving the micro-arc oxidation process, adding HA particles or rare earth, strontium, and copper elements into the electrolyte can change the surface morphology, resulting in the increase in hydrophilicity, which reduces the absorption tendency of protein, platelets, and coagulation factors. High surface roughness provides favorable attachment points for cell growth and migration, thus improving the blood compatibility of implanted materials. Moreover, compared with ordinary MAO coating, the coating containing HA has fewer cracks and pores, which further inhibits the penetration of the corrosive medium and has good corrosion resistance [50,51]. The combination of sliding technology (SFT), ultrasonic cold forging (UCFT), and powder coating double layer with micro-arc oxidation is effective because of the grain refinement of the coating. The deformed layer with plastic deformation was obtained by SFT treatment, and the grain size of the uppermost layer of the deformed layer was refined to nano-scale, which could make the cells show an extended polygonal shape, have strong adhesion with the coating, and show many filamentary pseudopods. Through UCFT treatment, the surface of magnesium alloy can be nano-sized, the pore diameter can be reduced, the number of pores can be increased, the compactness can be improved, and the defects of film can be reduced.

Although micro-arc oxidation can control pores by adding polymer sealing coating and combining with other methods, the high experimental cost is still an important reason why this method is difficult to be widely applied. Moreover, the oxidation voltage is much higher than that of conventional anodic oxidation, so safety protection measures should be taken during operation. During the experiment, the temperature of electrolyte rises rapidly, and large-capacity refrigeration and heat exchange equipment is required, which requires a large investment in the early stage.

5. Effect of Sol-Gel Method on Biocompatibility

Sol-gel method is a method in which liquid compounds with high chemical activity components undergo a series of chemical reactions to form gels and then oxidize into solids [52]. Sol-gel method is a general technology that has many advantages. (1) Because the raw materials are dispersed in the solvent, the reactants can be evenly mixed at the molecular level. When forming gel, it can be easily and evenly covered on the surface, especially suitable for irregular and complex surfaces. (2) Because of the step of solution reaction, it is easy to mix some trace elements uniformly and quantitatively, so as to achieve uniform doping at the molecular level. (3) Compared with the solid-phase reaction, the chemical reaction in solution is easier to carry out, and requires a lower synthesis temperature. It is generally believed that the diffusion of components in sol-gel system is in the nanometer range, whereas that in solid-phase reaction is in the micrometer range, so the purity of the formed coating is higher. Effective coatings with improved properties can be provided by changing the ratio of precursor to solvent, hydrolysis agent, curing temperature, hydrolysis rate, and deposition time [53]. There are two methods to prepare sol-gel: (a) inorganic and (b) organic. The inorganic method is to gel suspended colloidal particles with a particle size of 1–1000 nm to form a network. Nezamdoust et al. [54] used sol-gel coating containing different amounts of hydroxylated nano-diamond (HND) particles to treat magnesium alloy with corrosion resistance. Amaravathy et al. [55] synthesized niobium oxide (Nb_2O_5) coating on magnesium alloy by the sol-gel method. The surface characterization of the coating shows that the coating is composed of porous nanoparticles with a grain size of approximately 48 nm. Figure 5 shows the contrast of AZ31 with and without Nb_2O_5 coating. Niobium oxide coating has high microhardness and bonding strength, provides good surface anticorrosion protection, reduces degradation rate, and significantly enhances cell adhesion. Figure 6 shows the DAPI staining of osteoblast nuclei living on different magnesium alloy surfaces. Generally, the organic method is to dissolve metal monomer in organic solvent. Habib et al. [56] used tetraethoxysilane (TEOS) and methyltriethoxysilane (MTES) as raw materials to prepare sol-gel coating. The reason for the improvement of corrosion resistance is mainly the formation of a dense Si-O-Si network. Khramov et al. [57] prepared phosphate hybrid silane coating by mixing diethylphoxyethyl triethoxysilane and tetraethoxysilane in different molar ratios through hydrolysis and condensation. The enhancement of corrosion was due to the chemical reaction between phosphate and alloy surface, thus improving the hydrolysis stable P-O-Mg bond. In addition, the sol-gel method can not only combine with the matrix material at the molecular level to obtain uniform coating, but also be used as a sealing method for coatings with micropores and microcracks, such as MAO and CaP, to prepare mesoporous coatings [58,59]. Zhang et al. [60] prepared aminated hydroxyl cellulose (AHEC) coating on the surface after micro-arc oxidation by the sol-gel method, covering the pores of the micro-arc oxidation layer. Liu et al. [61] performed sol-gel post-treatment on the CaP coating to repair the pore gap of the coating.

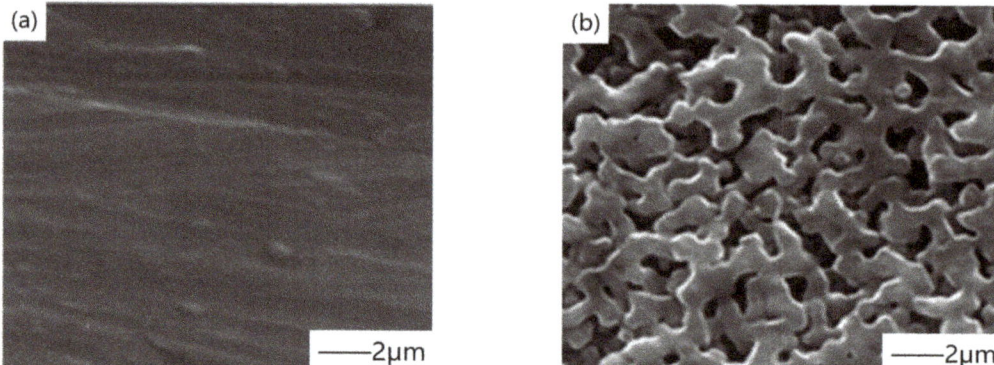

Figure 5. SEM analyses of (**a**) uncoated AZ31, (**b**) Nb2O5 coated AZ31 substrate. Reprinted with permission from Ref. [55]. Copyright 2014 Copyright Elsevier.

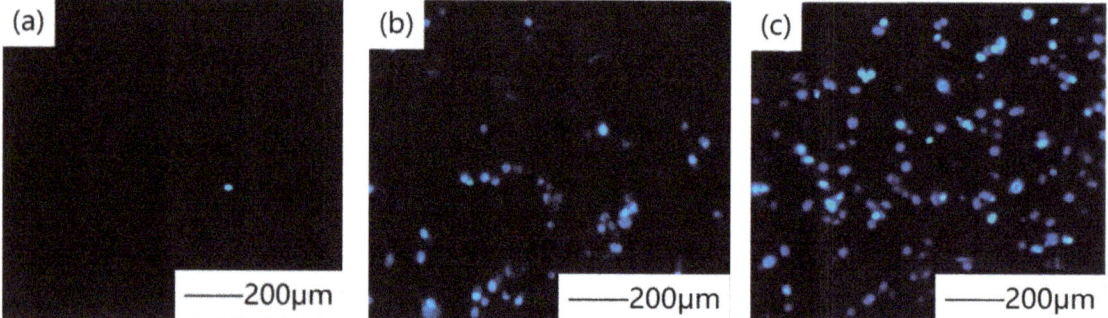

Figure 6. DAPI staining of nuclei of live osteoblast cells showing blue fluorescence on (**a**) uncoated and (**b**) Nb$_2$O$_5$ coated substrates after incubation with MG63 osteoblast cells for 24 h and (**c**) nuclei of live cells observed on Nb$_2$O$_5$ coated substrate after 48 h of incubation with MG63 osteoblast cells. (Ethidium bromide/acridine orange combined stain causes live cells to fluoresce green, whereas apoptotic cells cause the distinctive red-orange fluorescence.) Reprinted with permission from Ref. [55]. Copyright 2014 Copyright Elsevier.

The inorganic method can alleviate the corrosion because the hydroxide formed on the surface of the coating enhances the surface energy and improves the hydrophilicity through mechanical interlocking with ions. The nano-scale roughness of the coating is conducive to cell adhesion, porosity is conducive to bone integration, provides more contact surfaces for cell pseudopodia, and provides a way for cell nutrition penetration, so the morphology, proliferation rate, and expansion of cells are improved. The reason why the organic method can alleviate corrosion is that the formed silane-based sol-gel coating has low electrical reaction sensitivity with Mg, is easy to be chemically modified, and its chemical properties have low cytotoxicity to cells. Silicon and oxygen elements form a dense Si-O-Si network, which has stronger adhesion. In addition, the sealing moon, as a microporous and microcrack coating, can delay corrosion, because the permeable water of the sol condensation solvent diffuses, and the coating is preferentially attacked in the presence of water. At the same time, hydrogen generated by corrosion can be discharged through pores of the coating, and the hydrophilic porous structure on the surface promotes cell proliferation. Therefore, as a porous organic material, sol coagulation can not only act as a sealing layer, but can also provide good conditions for cell adhesion and proliferation [62]. However, the whole sol-gel process takes a long time, often several days or weeks. There are a lot of micropores in the gel, which will escape many gases and organic substances and cause shrinkage. The internal stress generated in the heat treatment process often leads to the cracking of the coating and reduces the corrosion resistance of the coating. Therefore, sol-gel coating technology needs further optimization, and how to control the coating thickness by changing parameters in the future to protect magnesium alloy matrix is a big development direction.

6. Effect of Electrophoretic Deposition on Biocompatibility

Electrophoretic deposition (EPD) is a process in which particles in suspension are deposited on the surface of a substrate using a DC electric field. The specific operation process generally means that under the action of the external DC electric field, the particles in the suspension move towards one end with opposite charges and gradually deposit a coating on the substrate material. The most critical factor affecting the surface morphology of the coating is suspension, including particle size, liquid dielectric constant, conductivity, medium viscosity, electromotive force, and stability of suspension [63]. In addition, deposition time, applied voltage, solid concentration in suspension, and substrate conductivity also have significant influence on the electrophoresis process [64]. Electrophoretic deposition has the advantage of forming a layer in a short time, requiring only simple equipment. In the application of electrophoretic deposition, hydroxyapatite coating is typically prepared. Antoniac et al. [65] prepared a uniform and continuous HAP coating with a thickness of approximately 15–16μm on the surface of Mg-Zn-Mn biodegradable alloys (ZMX410 and ZM21) by controlling the electrophoretic deposition parameters and compared the biodegradation rates before and after coating. Figure 7 shows the SEM micrographs of Mg alloy without coating and coated with HAP. The results show that the coating reduces the corrosion current density and improves the biocompatibility. This is because carbonation leads to the improvement of surface hydrophilicity and polarization resistance, which effectively improve the corrosion resistance of Mg-Zn-Mn alloy in a simulated environment. In addition, Wang et al. [66] prepared an HA layer with biological activity on MAO coating by the EPD method. MAO coating surface is loose and porous, which provides a good position for the deposition of HA particles, and HA particles will not easily fall off the sample surface, thus improving the corrosion resistance.

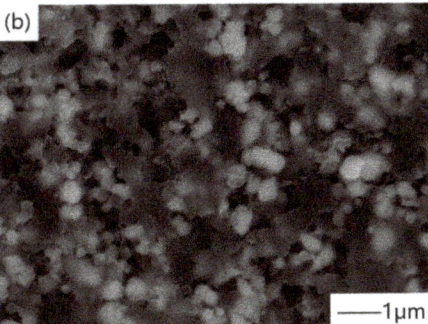

Figure 7. Comparison of SEM micrographs of Mg alloy coated with HAP and uncoated with HAP: (**a**) Mg alloy uncoated with HAP, (**b**) Mg alloy coated with HAP. Reprinted with permission from Ref. [65]. Copyright 2020 Copyright Elsevier.

Electrophoretic deposition has many advantages: (1) The preparation of HA coating is carried out in a mild environment, so there is no thermal stress at the interface between the substrate and the coating. This process can avoid the phase change and brittle fracture caused by high-temperature spraying, which is conducive to enhancing the bonding strength between the substrate and the coating. (2) Electrophoretic deposition is a non-linear process, so it is not limited by the shape and uniformity of the magnesium alloy surface. Ceramic coating can be prepared on the substrate with complex shape and porous surface, and the coating is plump, uniform, flat, and smooth. Electrophoresis has high dispersion ability, and even in the concave part of the product, a completely uniform protective film can be formed. (3) The thickness of the coating can vary in a large range. By accurately controlling the thickness and shape of the coating, it can vary from one micron to one hundred microns. The thickness of the coating can be controlled by adjusting different operating voltages to achieve extremely high corrosion resistance. (4) The deposition is fast, the yield is high, and it only takes a few seconds or minutes. The changeable preparation technology makes it have a broad commercial value. However, during electrophoretic deposition, HA particles are unevenly dispersed, and it is easy to form hydrogen bubbles at the magnesium surface, which seriously affects coating density and interfacial bonding strength. At the initial stage of electrophoresis, the investment is relatively large, and the workpiece that can be made can only be conductive. Of course, it can also be electrophoresed on plastic now, but a conductive layer has to be added. In the future, coatings with better biocompatibility and cell compatibility can be developed by combining with other methods to achieve better application in organisms.

7. Effect of Hydrothermal Method on Biocompatibility

Hydrothermal method refers to a method of preparing materials by dissolving and recrystallizing powder with water as solvent in a completely sealed pressure vessel. The main difference between the hydrothermal method and other wet chemical methods, such as the sol-gel method, lies in temperature and pressure. Hydrothermal methods usually use a temperature between 130 °C and 250 °C, and the corresponding water vapor pressure is 0.3 ~ 4 MPa. The coating prepared by hydrothermal method has the following advantages: (1) The coating can be directly obtained without high-temperature calcination, which avoids grain growth, defect formation, and impurity introduction during calcination, so the prepared coating has high sintering activity. (2) The prepared powder has the advantages of complete grain development, small particle size, uniform distribution, and light particle agglomeration, and it can be used as cheaper raw materials. The crystal structure, crystal morphology, and grain purity of nanoparticles can be controlled by adjusting the reaction conditions. The particle size range of the obtained powder materials is usually 0.1 μm to several microns, and some of them can reach tens of nanometers. Uniform coating can be prepared on surfaces with nonlinear and complex shapes. (3) The prepared calcium phosphate coating has high interfacial bonding strength and density, which is conducive to significantly improving the corrosion resistance of magnesium alloy biomaterials.

Compared with other technologies, the principle of preparing calcium phosphate (Ca-P) coating, especially hydroxyapatite coating (HAP), on the surface of magnesium alloy by hydrothermal treatment is relatively simple, and the operational device is easy to control. Therefore, more and more scholars use this method to carry out surface modification treatment on implantable biomedical magnesium alloy. Kang et al. [67] deposited superhydrophobic hydroxyapatite coating on magnesium alloy (Mg-Gd-1.5Nd-0.3Zn-0.3Zr) containing stearic acid and calcium phosphate compound by the one-step hydrothermal method. Figure 8 shows the SEM micrograph of superhydrophobic hydroxyapatite coating, and the test results show that the coating has low corrosion current density. This is because the superhydrophobic coating can make air stay in the layered structure of the coating, effectively reduce the contact between corrosive solution and the surface, and inhibit the exchange between corrosive ions and the surface. With the deepening of research, there have been many modified HAP in recent years. Zhou et al. [68] prepared a new type of hydroxyapatite coating which is denser than ordinary HA coating through the induction of poly dopamine, making it more corrosion-resistant. Peng et al. [69] enhanced the osteoinductivity of HA coating by adding zinc oxide nanoparticles with antibacterial activity. Yang et al. [70] introduced cationic surfactant tetradecyl trimethyl ammonium bromide (TTAB) for the first time and in-situ hydrothermally synthesized dense anti-corrosive magnesium–aluminum layered double hydroxide films on AZ31 magnesium alloy. In addition, hydrothermal method is often combined with chemical conversion method [71], micro-arc oxidation method [72], and sol coagulation method [73] to improve the coating performance.

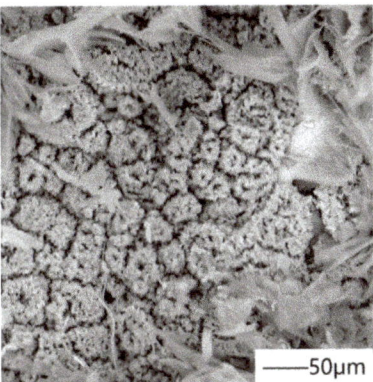

Figure 8. Shows the SEM micrograph of superhydrophobic hydroxyapatite coating. Reprinted with permission from Ref. [67]. Copyright 2018 Copyright Elsevier.

The core of the hydrothermal method is to select biocompatible materials with functional groups that coordinate with calcium ions, such as polydopamine, zinc oxide nanoparticles, amino acids, etc., which can chelate with calcium ions. The preparation of superhydrophobic coating can effectively improve the corrosion resistance of magnesium alloy, but the effect of superhydrophobic coating on the biological properties of magnesium alloy needs further study. In addition, the composite coating formed by other methods can also effectively improve the antibacterial and bone-guiding properties of calcium phosphate coatings, synthesize calcium phosphate coatings with different morphologies and phase compositions, significantly refine the grains of artificially synthesized HA crystals, and reduce their clustering effect [74]. It is worth noting that in order to protect the magnesium alloy coating for a long time, the density, thickness, solubility, and antibacterial ability of the coating must be comprehensively considered. These abilities of coatings are closely related to their chemical composition and micro-morphology, which directly affect the degradation rate. Therefore, it is a future development trend to design HA coating with comprehensive ability and strong bonding strength between coating and Mg matrix. However, the disadvantages of the hydrothermal method are obvious. The process is carried out in a closed container, and the growth process, which is not intuitive, cannot be observed. Moreover, it requires high equipment requirements, has great technical difficulty (strict control of temperature and pressure), and requires high-temperature and high-pressure steps, which makes it more dependent on production equipment. Therefore, the hydrothermal method tends to develop at low-temperature and low-pressure, that is, the temperature is lower than 100 °C and the pressure is close to one standard atmospheric pressure. Generally, only oxide powders can be prepared by the hydrothermal method, but there is a lack of in-depth research on the control of influencing factors of crystal nucleation and crystal growth, and no satisfactory conclusion has been reached.

8. Effect of Thermal Spraying Method

Thermal spraying is the use of high temperature and high speed to melt powder or metal wires as raw materials and deposit one material on the surface of another [75]. According to different heat sources, thermal spraying can be divided into flame thermal spraying, arc thermal spraying, plasma spraying, etc. Among them, plasma spraying (PSP) is the most common one in magnesium alloy surface modification. Plasma spraying systems consist of electric control power supply, plc-based operator control station, gas mass flow system, closed-loop water cooling system, powder feeder, and plasma torch.

The unique features of plasma spraying process are: (1) it can melt a variety of metals, ceramics, or composite materials [76,77]; (2) the deposition rate is high [78]; (3) the obtained coating is uniform [79]; and (4) the coating can be controlled by a variety of parameter settings [80]. Most importantly, the particle velocity of plasma spraying is higher than that of flame spraying and arc spraying, so the coating is denser and the surface morphology is finer [81].

In the preparation of medical magnesium alloy coating, spraying HA particles is the most common. Gao et al. [82] improve the corrosion resistance and bioactivity of magnesium alloy by plasma spraying hydroxyapatite coating. Figure 9 shows the surface morphology and cross-section microstructure of the coating. In recent years, with the deepening of research, everyone began to improve HA coating. Yao et al. [83] added zinc to hydroxyapatite. The corrosion rate of zinc was lower than that of magnesium, and zinc had a positive effect on bone mineralization and cell protein synthesis, thus improving the long-term viability and bone binding ability of the HA coating. Singh et al. [84] successfully sprayed Ta/Ti coating on AZ31B magnesium alloy by the plasma spraying process, which has high hardness and reduced wear rate. This behavior is attributed to the high passivation property of cold sprayed Ta/Ti coatings and the rough surface (island shape) of these coatings, which can provide proper nucleation sites for the formation and growth of calcium phosphate compounds in Hanks solution, and the dense structure of Ta layer hinders the penetration of corrosive solution. Mohajernia et al. [85] prepared a hydroxyapatite coating containing multi-walled carbon nanotubes (MWCNTs). Due to the high melting point of MWCNTs, they remained intact during plasma spraying and acted as a bridge between melted/semi-melted fragments, thus improving the fracture toughness of the HAP coating.

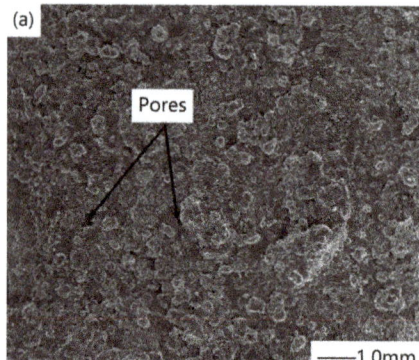
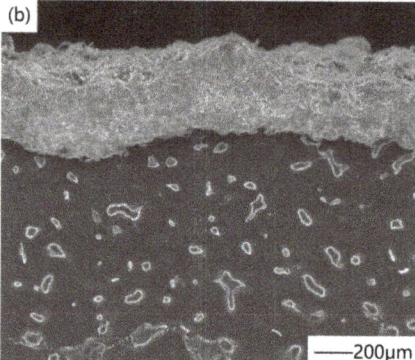

Figure 9. (a) Surface morphology of coating, (b) microstructure of coating section. Reprinted with permission from Ref. [82]. Copyright 2018 Copyright Elsevier.

Plasma spraying is mainly used in medical magnesium alloy to enhance its surface hardness. The grain size grows, the strengthening phase disappears, and the grain boundary strengthening phase grows, which is the fundamental reason for the hardness enhancement of the alloy. Moreover, the plasma spraying speed is very high, so the coating is denser and the surface morphology is finer. Specifically, during plasma spraying, molten HA particles hit magnesium alloy and rapidly solidified to form an amorphous phase, partially decomposed to form β-$Ca_3(PO_4)_2$, and the rest of the HA particles cooled to form HA phase, so the coating consists of amorphous phase, β-$Ca_3(PO_4)_2$, and HA phase. Compared with HA, amorphous phase and β-$Ca_3(PO_4)_2$ have higher solubility and degradation rate. The generated calcium and phosphorus ions react with protein molecules of bone cells, stimulating the growth of bone, and making implant materials form chemical-biological combination with bone, but at the same time, it also leads to the instability of the coating under the combined action of load and corrosion. The HA crystal is composed of unmelted or partially melted raw materials and recrystallized molten particles, which have a slow cooling rate, so they are not transformed into the amorphous phase. The dissolution rate of crystals in vivo and in vitro is slow, which is not conducive to bone growth, but it improves the mechanical properties of the coating and ensures the lasting bonding strength, which is particularly important at the interface between the coating and the implant. From the surface morphology of the coating, it can be seen from Figure 9 that the coating is composed of flat particles and a small amount of pores. The pores and rough structure of the plasma sprayed coating not only change the mechanical transmission mode of bone/implant, but also facilitate the climbing, growth, migration, and other activities of bone cells. For the development of HA coating in recent years, the modification is mainly aimed at improving the compactness of coating and reducing pore cracks. By adding Zn into HA coating by plasma spraying technology, ZnO clusters are formed on the surface of the coating, which enhances the bonding between particles, and the dense double-layer structure formed hinders the penetration corrosion of electrolyte to magnesium alloy substrate. According to the difference of thermal conductivity between HA and Ta, adding Ta can adjust the crystallinity of the coating, make the surface relatively flat, reduce the fused solidified particles, and improve the protective ability of the surface. Multi-walled carbon nanotubes (MWNTs) can improve the fracture toughness of the coating and limit the crack propagation, because they act as a bridge, and more energy is required to destroy it. Moreover, the existence of MWNTs will reduce the content of the secondary phase, thus affecting the fracture mechanism. However, the HA coating formed by plasma spraying has rough micro-scale, pores, and cracks, and poor bone conductivity. The high temperature in the processing leads to the decomposition of the calcium phosphate phase and the formation of the amorphous phase, and the purity and crystallinity of the coating are low, which accelerates the dissolution of the HA coating, leading to its loss of synchronization with the life of the implant. As far as the development of plasma spraying metal-based ceramic coatings is concerned, in order to further improve the density of coatings and reduce the cost of plasma spraying ceramic coatings, efforts should be made in the following aspects. First, optimize the coating composition and process parameters, and combine with more methods. Second, according to the requirements of different parts, a new type of environmentally friendly hole sealing agent was developed. Finally, while meeting different requirements, the production equipment is optimized and the production cost is reduced. Among many preparation methods of nano-ceramic coatings, thermal spraying technology is the most likely to produce market benefits in a short time.

9. Summary and Prospect

In this paper, six surface modification methods—chemical conversion method, micro-arc oxidation method, sol-gel method, electrophoretic deposition, hydrothermal method, and thermal spraying method—are reviewed. Chemical conversion method has good binding force because its own material is converted into coating. Micro-arc oxidation method combines with sol-gel method, which is often used as sealing layer, to obtain better coating. Electrophoretic deposition is not limited by the shape of magnesium alloy and can quickly form uniform and smooth coating. Because the hydrothermal method does not need high-temperature calcination treatment, the coating has high sintering activity, and thermal spraying improves metallographic structure and corrosion resistance by melting various materials. To compare the differences of the six methods more conveniently, their definitions, advantages and disadvantages, characteristics, and the number of literature publications are summarized in Table 1. Because magnesium alloy is implanted into the body as a bone plate, it must have strong mechanical properties, good corrosion resistance, and biocompatibility. At present, the most studied method is coating priming by micro-arc oxidation and then post-treatment by sol-gel method. However, the cost of this method is much higher than that of hydrothermal method and thermal spraying method. Considering that the coating of the implant should not fall off and be uniform, chemical conversion method and electrophoretic deposition can solve this problem well. In order to understand the development of the six methods, the number of publications was searched in the Web of Science database. There were 992 articles on the chemical conversion method, and the most published in 2019 was about 120. Micro-arc oxidation method (693) published the most in 2019, about 80 articles. Sol method (559) published the most in 2018, about 60 articles. Electrophoretic deposition method (618) published the most in 2019, about 80 articles. Hydrothermal method (394) published the most in 2021, with about 65 articles. Thermal spraying method (188) published the most in 2019, about 30 articles. However, these technical methods are not very mature in three aspects. (1) Biological safety: At present, most of the research on medical coated magnesium alloy is in vitro research, and only a few alloys are implanted into animals. However, the actual internal environment of the human body is more complicated, and it is impossible to accurately grasp the problems such as excessive hydrogen evolution, inflammation, and degradation of metal elements poisoning, and there is still a long way to go before clinical application. (2) Surface properties: Medical coated magnesium alloy has different functions due to different implantation sites. For example, the surface morphology of magnesium alloy bone plate needs to be conducive to the growth of cells, and the surface of magnesium alloy vascular support tube needs antithrombotic performance, etc. At present, although there are many methods to form coatings, it is difficult for the morphology formed on the surface to reach the ideal state. (3) Mechanical properties: At present, the research of medical coated magnesium alloy is mainly used as bone screws, but a single coating cannot provide stable degradation protection and cannot achieve the purpose of effectively connecting broken bones.

Table 1. Comparison of modification methods of medical magnesium alloy surface coating.

Method	Chemical Conversion	Microarc Oxidation	Sol Coagulation	Electrophoretic Deposition	Hydrothermal Method	Thermal Spraying
Concept	The material itself is transformed into a coating through chemical changes.	High-voltage plasma assisted anodic oxidation process.	The liquid is converted into gel and then oxidized into a solid.	Particles are deposited on the surface of the substrate by direct current electric field.	With water as solvent, the powder is dissolved and recrystallized.	The powder is deposited on the surface at high temperature and high speed.
Adv.	The coating has good binding force and low cost.	With high hardness, it is often used as the base coat of the first step.	Uniform film formation, often used as the second step sealing coating.	Uniform coat can be prepared on a substrate with complex shape.	Avoids impurity entry and defect growth in that calcination process.	High deposition rate, uniform coating.
Disad.	Porous, cracked, and pollutes the environment.	Characteristic micropores make it corrosion-resistant.	Results in thermal stress cracking of the coating.	Large investment, difficult process control.	High equipment requirements and technical difficulties.	Poor bonding strength and environmental pollution.
Feature	Self-transformation, good binding force.	Coating with high hardness and abrasion resistance.	As the second step of sealing coating.	Avoid embrittlement caused by high temperature.	Coating has high sintering activity.	Improving metallographic structure and corrosion resistance.
Number	992	693	559	618	394	188

In view of the above problems, the development trend of medical coated magnesium alloy in the future is as follows: (1) Mastering the degradation mechanism of medical coated magnesium alloy through in vitro research and animal in vivo research, and studying the degradation conditions in human body through corrosion simulation, the coating can avoid using metal elements that may lead to poisoning, and can alleviate the precipitation of hydrogen. Coating magnesium alloy for medical use will reduce the cost on the basis of green production, and it will be widely used in clinical settings. (2) To improve the pertinence of magnesium alloy coating, various methods can be used at the same time to form ideal coating with pore size, surface roughness, hydrophilicity, and hydrophobicity. (3) Through the simulation of corrosion mechanics, the stress situation of each part is judged, and the mechanical properties (such as hardness, strength, etc.) are enhanced by multi-coating superposition in places with large stress. Multi-coating superposition is the development trend in the future, which makes medical coated magnesium alloy widely used for flexible coating. In a word, surface coating modification has created the possibility for further application of magnesium alloy in the medical field. It is believed that with the continuous development of surface coating modification technology of magnesium alloy, magnesium alloy will play a greater role in the field of biomedical valuable materials.

Author Contributions: Conceptualization, Y.Q. and X.G.; methodology, X.G.; software, Y.H.; validation, Y.Q., Y.H. and K.Y.; formal analysis, K.Y.; investigation, K.Y.; resources, Y.Q.; data curation, Y.H.; writing—original draft preparation, X.G.; writing—review and editing, X.G.; visualization, K.Y.; supervision, K.Y.; project administration, Y.H.; funding acquisition, Y.H. All authors have read and agreed to the published version of the manuscript.

Funding: This research was funded by Natural Science Foundation of Shandong Province (ZR2019-QEE032), the Shandong Higher Education Youth Innovation and Technology Support Program (2019KJB021), the Independent Innovation Team Foundation of Jinan (2019GXRC012).

Institutional Review Board Statement: Not applicable.

Informed Consent Statement: Not applicable.

Data Availability Statement: Not applicable.

Conflicts of Interest: The authors declare no conflict of interest.

References

1. Farraro, F.K.; Kim, E.K.; Woo, L.S.; Flowers, R.J.; McCullough, B.M. Revolutionizing orthopaedic biomaterials: The potential of biodegradable and bioresorbable magnesium-based materials for functional tissue engineering. *J. Biomech.* **2014**, *47*, 1979–1986. [CrossRef] [PubMed]
2. Alaneme, K.K.; Okotete, A.E. Enhancing plastic deformability of Mg and its alloys—A review of traditional and nascent developments. *J. Magnes. Alloy.* **2017**, *5*, 460–475. [CrossRef]
3. Atrens, A.; Song, G.L.; Cao, F.; Shi, Z.; Bowen, P.K. Advances in Mg corrosion and research suggestions. *J. Magnes. Alloy.* **2013**, *1*, 177–200. [CrossRef]
4. Moghaddam, N.S.; Andani, M.T.; Amerinatanzi, A.; Haberland, C.; Huff, S.; Miller, M.; Dean, D. Metals for bone implants: Safety, design, and efficacy. *Biomanuf. Rev.* **2016**, *1*, 1–16. [CrossRef]
5. Sumner, D.R. Long-term implant fixation and stress-shielding in total hip replacement. *J. Biomech.* **2015**, *48*, 797–800. [CrossRef] [PubMed]
6. Atrens, A.; Johnston, S.; Shi, Z.; Dargusch, M.S. Understanding Mg corrosion in the body for biodegradable medical implants. *Scr. Mater.* **2018**, *154*, 92–100. [CrossRef]
7. Hermawan, H.; Dubé, D.; Mantovani, D. Degradable metallic biomaterials: Design and development of Fe-Mn alloys for stents. *J. Biomed. Mater. Res. Part A* **2010**, *93*, 1–11. [CrossRef]
8. Bairagi, D.; Mandal, S. A comprehensive review on biocompatible Mg-based alloys as temporary orthopaedic implants: Current status, challenges, and future prospects. *J. Magnes. Alloy.* **2021**, *10*, 627–669. [CrossRef]
9. Razavi, M.; Huang, Y. Assessment of magnesium-based biomaterials: From bench to clinic. *Biomater. Sci.* **2019**, *7*, 2241–2263. [CrossRef]
10. Surmeneva, M.A.; Vladescu, A.; Cotrut, C.M.; Tyurin, A.; Pirozhkova, T.; Shuvarin, I.; Elkin, B.; Oehr, C.; Surmenev, R. Effect of parylene C coating on the antibiocorrosive and mechanical properties of different magnesium alloys. *Appl. Surf. Sci.* **2018**, *427*, 617–627. [CrossRef]
11. Elkamel, R.S.; Fekry, A.M.; Ghoneim, A.A.; Filippov, L.O. Electrochemical Corrosion Behaviour of AZ91E Magnesium Alloy by means of Various Nanocoatings in Aqueous Peritoneal Solution: In Vitro and In Vivo Studies. *J. Mater. Res. Technol.* **2022**, *17*, 828–839. [CrossRef]
12. El-Kamel, R.S.; Ghoneim, A.A.; Fekry, A.M. Electrochemical, biodegradation and cytotoxicity of graphene oxide nanoparticles/polythreonine as a novel nano-coating on AZ91E Mg alloy staple in gastrectomy surgery. *Mater. Sci. Eng. C* **2019**, *103*, 109780–109789. [CrossRef] [PubMed]
13. Prabhu, D.B.; Gopalakrishnan, P.; Ravi, K.R. Morphological studies on the development of chemical conversion coating on surface of Mg-4Zn alloy and its corrosion and bio mineralisation behaviour in simulated body fluid. *J. Alloys Compd.* **2020**, *812*, 152146. [CrossRef]
14. Devgan, S.; Sidhu, S.S. Evolution of surface modification trends in bone related biomaterials: A review. *Mater. Chem. Phys.* **2019**, *233*, 68–78. [CrossRef]
15. Li, J.A.; Chen, L.; Zhang, X.Q.; Guan, S.K. Enhancing biocompatibility and corrosion resistance of biodegradable Mg-Zn-Y-Nd alloy by preparing PDA/HA coating for potential application of cardiovascular biomaterials. *Mater. Sci. Eng. C* **2020**, *109*, 110607–110620. [CrossRef]

16. Huynh, V.; Ngo, N.K.; Golden, T.D. Surface activation and pretreatments for biocompatible metals and alloys used in biomedical applications. *Int. J. Biomater.* **2019**, *2019*, 1–21. [CrossRef]
17. Surmeneva, M.A.; Ivanova, A.A.; Tian, Q.; Pittman, R.; Jiang, W.; Lin, J.; Liu, H.H.; Surmenev, R.A. Bone marrow derived mesenchymal stem cell response to the RF magnetron sputter deposited hydroxyapatite coating on AZ91 magnesium alloy. *Mater. Chem. Phys.* **2019**, *221*, 89–98. [CrossRef]
18. Surmeneva, M.A.; Mukhametkaliyev, T.M.; Khakbaz, H.; Surmenev, R.A.; Kannan, M.B. Ultrathin film coating of hydroxyapatite (HA) on a magnesium–calcium alloy using RF magnetron sputtering for bioimplant applications. *Mater. Lett.* **2015**, *152*, 280–282. [CrossRef]
19. Tian, L.; Tang, N.; Ngai, T.; Wu, C.; Ruan, Y.; Huang, L.; Qin, L. Hybrid fracture fixation systems developed for orthopaedic applications: A general review. *J. Orthop. Transl.* **2019**, *16*, 1–13. [CrossRef]
20. Chandra, G.; Pandey, A. Biodegradable bone implants in orthopedic applications: A review. *Biocybern. Biomed. Eng.* **2020**, *40*, 596–610. [CrossRef]
21. Wan, P.; Tan, L.; Yang, K. Surface modification on biodegradable magnesium alloys as orthopedic implant materials to improve the bio-adaptability: A review. *J. Mater. Sci. Technol.* **2016**, *32*, 827–834. [CrossRef]
22. Hornberger, H.; Virtanen, S.; Boccaccini, A.R. Biomedical coatings on magnesium alloys-A review. *Acta Biomater.* **2012**, *8*, 2442–2455. [CrossRef] [PubMed]
23. Chen, X.B.; Birbilis, N.; Abbott, T.B. Review of Corrosion-Resistant Conversion Coatings for Magnesium and Its Alloys. *Corrosion* **2011**, *67*, 035005. [CrossRef]
24. Xu, L.P.; Pan, F.; Yu, G.; Yang, L.; Zhang, E.; Yang, K. In Vitro and In Vivo evaluation of the surface bioactivity of a calcium phosphate coated magnesium alloy. *Biomaterials* **2009**, *30*, 1512–1523. [CrossRef] [PubMed]
25. Carla, L.; Johannes, G.B.; Philip, K. Effect of surface pre-treatments on biocompatibility of magnesium. *Acta Biomater.* **2009**, *5*, 2783–2789. [CrossRef]
26. Yang, J.X.; Cui, F.Z.; Lee, I.S.; Zhang, Y.; Yin, Q.; Xia, H.; Yang, S. In Vivo biocompatibility and degradation behavior of Mg alloy coated by calcium phosphate in a rabbit model. *J. Biomater. Appl.* **2012**, *27*, 153–164. [CrossRef]
27. Carboneras, M.; Garcia-Alonso, M.C.; Escudero, M.L. Biodegradation kinetics of modified magnesium-based materials in cell culture medium. *Corros. Sci.* **2011**, *53*, 1433–1439. [CrossRef]
28. Chiu, K.Y.; Wong, M.H.; Cheng, F.T. Characterization and corrosion studies of fluoride conversion coating on degradable Mg implants. *Surf. Coat. Technol.* **2007**, *202*, 590–598. [CrossRef]
29. Witte, F.; Fischer, I.J.; Nellesen, J.; Vogt, C.; Vogt, J.; Donath, T.; Beckmann, F. In Vivo corrosion and corrosion protection of magnesium alloy LAE442. *Acta Biomater.* **2010**, *6*, 1792–1799. [CrossRef]
30. Zhu, Y.; Gao, W.; Huang, H.; Chang, W.; Zhang, S.; Zhang, R.; Zhang, Y. Investigation of corrosion resistance and formation mechanism of calcium-containing coatings on AZ31B magnesium alloy. *Appl. Surf. Sci.* **2019**, *487*, 581–592. [CrossRef]
31. Shin, K.R.; Ko, Y.G.; Shin, D.H. Surface characteristics of ZrO_2-containing oxide layer in titanium by plasma electrolytic oxidation in $K_4P_2O_7$ electrolyte. *J. Alloys Compd.* **2012**, *536*, S226–S230. [CrossRef]
32. Yigit, O.; Ozdemir, N.; Dikici, B.; Kaseem, M. Surface properties of graphene functionalized TiO_2/nHA hybrid coatings made on Ti6Al7Nb alloys via plasma electrolytic oxidation (PEO). *Molecules* **2021**, *26*, 3903. [CrossRef] [PubMed]
33. Zehra, T.; Kaseem, M.; Hossain, S.; Ko, Y.G. Fabrication of a protective hybrid coating composed of TiO_2, MoO_2, and SiO_2 by plasma electrolytic oxidation of titanium. *Metals* **2021**, *11*, 1182. [CrossRef]
34. Lu, J.P.; Cao, G.P.; Quan, G.F.; Wang, C.; Zhuang, J.J.; Song, R.G. Effects of voltage on microstructure and corrosion resistance of micro-arc oxidation ceramic coatings formed on KBM10 magnesium alloy. *J. Mater. Eng. Perform.* **2018**, *27*, 147–154. [CrossRef]
35. Ezhilselvi, V.; Nithin, J.; Balaraju, J.N.; Subramanian, S. The influence of current density on the morphology and corrosion properties of MAO coatings on AZ31B magnesium alloy. *Surf. Coat. Technol.* **2016**, *288*, 221–229. [CrossRef]
36. Yu, H.; Dong, Q.; Dou, J.; Pan, Y.; Chen, C. Preparation of Si-containing oxide coating and biomimetic apatite induction on magnesium alloy. *Appl. Surf. Sci.* **2016**, *388*, 148–154. [CrossRef]
37. Yong, J.; Li, H.; Li, Z.; Chen, Y.; Wang, Y.; Geng, J. Effect of $(NH_4)_2ZrF_6$, voltage and treating time on corrosion resistance of micro-arc oxidation coatings applied on ZK61M magnesium alloys. *Materials* **2021**, *14*, 7410. [CrossRef]
38. Yao, J.T.; Wang, S.; Zhou, Y.; Dong, H. Effects of the power supply mode and loading parameters on the characteristics of micro-arc oxidation coatings on magnesium alloy. *Metals* **2020**, *10*, 1452. [CrossRef]
39. Xiong, Y.; Yang, Z.; Hu, X.; Song, R. Bioceramic coating produced on AZ80 magnesium alloy by one-step microarc oxidation process. *J. Mater. Eng. Perform.* **2019**, *28*, 1719–1727. [CrossRef]
40. Yu, H.; Dong, Q.; Dou, J.; Pan, Y.; Chen, C. Structure and in vitro bioactivity of ceramic coatings on magnesium alloys by microarc oxidation. *Appl. Surf. Sci.* **2016**, *388*, 114–119. [CrossRef]
41. Han, L.; Li, X.; Xue, F.; Chu, C.; Bai, J. Biocorrosion behavior of micro-arc-oxidized AZ31 magnesium alloy in different simulated dynamic physiological environments. *Surf. Coat. Technol.* **2019**, *361*, 240–248. [CrossRef]
42. Ding, H.Y.; Li, H.; Wang, G.Q.; Liu, T.; Zhou, G.H. Bio-Corrosion Behavior of Ceramic Coatings Containing Hydroxyapatite on Mg-Zn-Ca Magnesium Alloy. *Appl. Sci.* **2018**, *8*, 569. [CrossRef]
43. Jianzhong, L.; Yanwen, T.; Zuoxing, G.U.I.; Huang, Z. Effects of rare earths on the microarc oxidation of a magnesium alloy. *Rare Met.* **2008**, *27*, 50–54. [CrossRef]

44. Sedelnikova, M.B.; Sharkeev, Y.P.; Tolkacheva, T.V.; Khimich, M.A.; Bakina, O.V.; Fomenko, A.N.; Epple, M. Comparative Study of the structure, properties, and corrosion behavior of Sr-containing biocoatings on Mg0.8Ca. *Materials* **2020**, *13*, 1942. [CrossRef]
45. Lee, H.P.; Lin, D.J.; Yeh, M.L. Phenolic modified ceramic coating on biodegradable Mg alloy: The improved corrosion resistance and osteoblast-like cell activity. *Materials* **2017**, *10*, 696. [CrossRef]
46. Ahmed, M.; Qi, Y.; Zhang, L.; Yang, Y.; Abas, A.; Liang, J.; Cao, B. Influence of Cu^{2+} ions on the corrosion resistance of AZ31 magnesium alloy with microarc oxidation. *Materials* **2020**, *13*, 2647. [CrossRef]
47. Huo, W.; Lin, X.; Lv, L.; Cao, H.; Yu, S.; Yu, Z.; Zhang, Y. Manipulating the degradation behavior and biocompatibility of Mg alloy through a two-step treatment combining sliding friction treatment and micro-arc oxidation. *J. Mater. Chem. B* **2018**, *6*, 6431–6443. [CrossRef]
48. Yang, J.; Gu, Y.; Zhou, X.; Zhang, Y. Tribocorrosion behavior and mechanism of micro-arc oxidation Ca/P coating on nanocrystallized magnesium alloys. *Mater. Corros.* **2018**, *69*, 749–759. [CrossRef]
49. Ballam, L.R.; Arab, H.; Bestetti, M.; Franz, S.; Masi, G.; Sola, R.; Martini, C. Improving the corrosion resistance of wrought ZM21 magnesium alloys by plasma electrolytic oxidation and powder coating. *Materials* **2021**, *14*, 2268. [CrossRef]
50. Kim, K.; Yu, M.; Zong, X.; Fang, D.; Seo, Y.-S.; Hsiao, B.S.; Chu, B.; Hadjiargyrou, M. Control of degradation rate and hydrophilicity in electrospun non-woven poly(d,l-lactide) nanofiber scaffolds for biomedical applications. *Biomaterials* **2003**, *24*, 4977–4985. [CrossRef]
51. Soria, J.M.; Ramos, C.M.; Bahamonde, O.; Cruz, D.M.G.; Sánchez, M.S.; Esparza, M.A.G.; Casas, C.; Guzmán, M.; Navarro, X.; Ribelles, J.L.G.; et al. Influence of the substrate's hydrophilicity on the in vitro Schwann cells viability. *J. Biomed. Mater. Res.* **2007**, *83*, 463–470. [CrossRef] [PubMed]
52. Talha, M.; Ma, Y.; Xu, P.; Wang, Q.; Lin, Y.; Kong, X. Recent advancements in corrosion protection of magnesium alloys by silane-based sol–gel coatings. *Ind. Eng. Chem. Res.* **2020**, *59*, 19840–19857. [CrossRef]
53. Qiu, Z.; Yin, B.; Wang, J.; Sun, J.; Tong, Y.; Li, L.; Wang, R. Theoretical and experimental studies of sol–gel electrodeposition on magnesium alloy. *Surf. Interface Anal.* **2021**, *53*, 432–439. [CrossRef]
54. Nezamdoust, S.; Seifzadeh, D.; Habibi-Yangjeh, A. Nanodiamond incorporated sol–gel coating for corrosion protection of magnesium alloy. *Trans. Nonferrous Met. Soc. China* **2020**, *30*, 1535–1549. [CrossRef]
55. Amaravathy, P.; Sowndarya, S.; Sathyanarayanan, S.; Rajendran, N. Novel sol gel coating of Nb_2O_5 on magnesium alloy for biomedical applications. *Surf. Coat. Technol.* **2014**, *244*, 131–141. [CrossRef]
56. Ashassi-Sorkhabi, H.; Moradi-Alavian, S.; Kazempour, A. Salt-nanoparticle systems incorporated into sol-gel coatings for corrosion protection of AZ91 magnesium alloy. *Prog. Org. Coat.* **2019**, *135*, 475–482. [CrossRef]
57. Khramov, A.N.; Balbyshev, V.N.; Kasten, L.S.; Mantz, R.A. Sol–gel coatings with phosphonate functionalities for surface modification of magnesium alloys. *Thin Solid Films* **2006**, *514*, 174–181. [CrossRef]
58. Nezamdoust, S.; Seifzadeh, D.; Rajabalizadeh, Z. Application of novel sol–gel composites on magnesium alloy. *J. Magnes. Alloy.* **2019**, *7*, 419–432. [CrossRef]
59. Pezzato, L.; Rigon, M.; Martucci, A.; Brunelli, K.; Dabalà, M. Plasma Electrolytic Oxidation (PEO) as pre-treatment for sol–gel coating on aluminum and magnesium alloys. *Surf. Coat. Technol.* **2019**, *366*, 114–123. [CrossRef]
60. Zhang, L.; Wu, Y.; Zeng, T.; Wei, Y.; Zhang, G.; Liang, J.; Cao, B. Preparation and characterization of a sol–gel ahec pore-sealing film prepared on micro arc oxidized AZ31 magnesium alloy. *Metals* **2021**, *11*, 784. [CrossRef]
61. Liu, Y.; Cheng, X.; Wang, X.; Sun, Q.; Wang, C.; Di, P.; Lin, Y. Micro-arc oxidation-assisted sol–gel preparation of calcium metaphosphate coatings on magnesium alloys for bone repair. *Mater. Sci. Eng. C* **2021**, *131*, 112491–112501. [CrossRef] [PubMed]
62. Weng, W.; Wu, W.; Yu, X.; Sun, M.; Lin, Z.; Ibrahim, M.; Yang, H. Effect of gel MA hydrogel coatings on corrosion resistance and biocompatibility of MAO-coated Mg alloys. *Materials* **2020**, *13*, 3834. [CrossRef] [PubMed]
63. Akram, M.; Arshad, N.; Aktan, M.K.; Braem, A. Alternating current electrophoretic deposition of chitosan–gelatin–bioactive glass on Mg–Si–Sr alloy for corrosion protection. *ACS Appl. Bio Mater.* **2020**, *3*, 7052–7060. [CrossRef] [PubMed]
64. Maqsood, M.F.; Raza, M.A.; Ghauri, F.A.; Rehman, Z.U.; Ilyas, M.T. Corrosion study of graphene oxide coatings on AZ31B magnesium alloy. *J. Coat. Technol. Res.* **2020**, *17*, 1321–1329. [CrossRef]
65. Antonioc, I.; Miculescu, F.; Cotrut, C. Controlling the degradation rate of biodegradable Mg–Zn–Mn alloys for orthopedic applications by electrophoretic deposition of hydroxyapatite coating. *Materials* **2020**, *13*, 263. [CrossRef]
66. Wang, Z.X.; Xu, L.; Zhang, J.W.; Ye, F.; Lv, W.G.; Xu, C.; Yang, J. Preparation and degradation behavior of composite bio-coating on ZK60 magnesium alloy using combined micro-arc oxidation and electrophoresis deposition. *Front. Mater.* **2020**, *7*, 190–204. [CrossRef]
67. Kang, Z.X.; Zhang, J.Y.; Niu, L. A one-step hydrothermal process to fabricate superhydrophobic hydroxyapatite coatings and determination of their properties. *Surf. Coat. Technol.* **2018**, *334*, 84–89. [CrossRef]
68. Zhou, Z.; Zheng, B.; Gu, Y.; Shen, C.; Wen, J.; Meng, Z.; Qin, A. New approach for improving anticorrosion and biocompatibility of magnesium alloys via polydopamine intermediate layer-induced hydroxyapatite coating. *Surf. Interfaces* **2020**, *19*, 100501–100518. [CrossRef]
69. Peng, M.; Hu, F.; Du, M.; Mai, B.; Zheng, S.; Liu, P.; Chen, Y. Hydrothermal growth of hydroxyapatite and ZnO bilayered nanoarrays on magnesium alloy surface with antibacterial activities. *Front. Mater. Sci.* **2020**, *14*, 14–23. [CrossRef]
70. Yang, Q.; Tabish, M.; Wang, J.; Zhao, J. Enhanced corrosion resistance of layered double hydroxide films on Mg alloy: The key role of cationic surfactant. *Materials* **2022**, *15*, 2028. [CrossRef]

71. Yuan, J.; Yuan, R.; Wang, J.; Li, Q.; Xing, X.; Liu, X.; Hu, W. Fabrication and corrosion resistance of phosphate/ZnO multilayer protective coating on magnesium alloy. *Surf. Coat. Technol.* **2018**, *352*, 74–83. [CrossRef]
72. Zhu, J.; Jia, H.; Liao, K.; Li, X. Improvement on corrosion resistance of micro-arc oxidized AZ91D magnesium alloy by a pore-sealing coating. *J. Alloys Compd.* **2021**, *889*, 161460–161469. [CrossRef]
73. Liu, W.; Yan, Z.; Ma, X.; Geng, T.; Wu, H.; Li, Z. Mg-MOF-74/MgF$_2$ composite coating for improving the properties of magnesium alloy implants: Hydrophilicity and corrosion resistance. *Materials* **2018**, *11*, 396. [CrossRef] [PubMed]
74. Sun, J.; Cai, S.; Wei, J.; Ling, R.; Liu, J.; Xu, G. Long-term corrosion resistance and fast mineralization behavior of micro-nano hydroxyapatite coated magnesium alloy In Vitro. *Ceram. Int.* **2020**, *46*, 824–832. [CrossRef]
75. Berndt, C.C.; Hasan, F.; Tietz, U.; Schmitz, K.P. A review of hydroxyapatite coatings manufactured by thermal spray. *Adv. Calcium Phosphate Biomater.* **2014**, *2*, 267–329. [CrossRef]
76. Bansal, P.; Singh, G.; Sidhu, H.S. Investigation of surface properties and corrosion behavior of plasma sprayed HA/ZnO coatings prepared on AZ31 Mg alloy. *Surf. Coat. Technol.* **2020**, *401*, 126241–126261. [CrossRef]
77. Mardali, M.; SalimiJazi, H.R.; Karimzadeh, F.; Luthringer, B.; Blawert, C.; Labbaf, S. Comparative study on microstructure and corrosion behavior of nanostructured hydroxyapatite coatings deposited by high velocity oxygen fuel and flame spraying on AZ61 magnesium based substrates. *Appl. Surf. Sci.* **2019**, *465*, 614–624. [CrossRef]
78. Yao, H.L.; Hu, X.Z.; Wang, H.T.; Chen, Q.Y.; Bai, X.B.; Zhang, M.X.; Ji, G.C. Microstructure and corrosion behavior of thermal-sprayed hydroxyapatite/magnesium composite coating on the surface of AZ91D magnesium alloy. *J. Therm. Spray Technol.* **2019**, *28*, 495–503. [CrossRef]
79. Daroonparvar, M.; Khan, M.F.; Saadeh, Y.; Kay, C.M.; Gupta, R.K.; Kasar, A.K.; Bakhsheshi-Rad, H.R. Enhanced corrosion resistance and surface bioactivity of AZ31B Mg alloy by high pressure cold sprayed monolayer Ti and bilayer Ta/Ti coatings in simulated body fluid. *Mater. Chem. Phys.* **2020**, *256*, 123627. [CrossRef]
80. Wang, Q.; Sun, Q.; Zhang, M.X.; Niu, W.J.; Tang, C.B.; Wang, K.S.; Wang, L. The influence of cold and detonation thermal spraying processes on the microstructure and properties of Al-based composite coatings on Mg alloy. *Surf. Coat. Technol.* **2018**, *352*, 627–633. [CrossRef]
81. García-Rodríguez, S.; López, A.J.; Bonache, V.; Torres, B.; Rams, J. Fabrication, Wear, and Corrosion Resistance of HVOF Sprayed WC-12Co on ZE41 Magnesium Alloy. *Coatings* **2020**, *10*, 502. [CrossRef]
82. Gao, Y.L.; Liu, Y.; Song, X.Y. Plasma-sprayed hydroxyapatite coating for improved corrosion resistance and bioactivity of magnesium alloy. *J. Therm. Spray Technol.* **2018**, *27*, 1381–1387. [CrossRef]
83. Yao, H.L.; Yi, Z.H.; Yao, C.; Zhang, M.X.; Wang, H.T.; Li, S.B.; Ji, G.C. Improved corrosion resistance of AZ91D magnesium alloy coated by novel cold-sprayed Zn-HA/Zn double-layer coatings. *Ceram. Int.* **2020**, *46*, 7687–7693. [CrossRef]
84. Singh, B.; Singh, G.; Sidhu, B.S. Analysis of corrosion behaviour and surface properties of plasma-sprayed composite coating of hydroxyapatite–tantalum on biodegradable Mg alloy ZK60. *J. Compos. Mater.* **2019**, *53*, 2661–2673. [CrossRef]
85. Mohajernia, S.; Pour-Ali, S.; Hejazi, S.; Saremi, M.; Kiani-Rashid, A.R. Hydroxyapatite coating containing multi-walled carbon nanotubes on AZ31 magnesium: Mechanical-electrochemical degradation in a physiological environment. *Ceram. Int.* **2018**, *44*, 8297–8305. [CrossRef]

Article

Microstructures and Properties of Al-Mg Alloys Manufactured by WAAM-CMT

Yan Liu [1,*], Zhaozhen Liu [2], Guishen Zhou [2], Chunlin He [1] and Jun Zhang [3]

1. Liaoning Provincial Key Laboratory of Advanced Material Preparation Technology, Shenyang University, Shenyang 110044, China
2. School of Mechanical Engineering, Shenyang University, Shenyang 110044, China
3. Liaoning Provincial Key Laboratory of Research and Application of Multiple Hard Films, Shenyang University, Shenyang 110044, China
* Correspondence: liuyan1979@syu.edu.cn; Tel.: +86-024-6226-9467

Abstract: A wire arc additive manufacturing system, based on cold metal transfer technology, was utilized to manufacture the Al-Mg alloy walls. ER5556 wire was used as the filler metal to deposit Al-Mg alloys layer by layer. Based on the orthogonal experiments, the process parameters of the welding current, welding speed and gas flow, as well as interlayer residence time, were adjusted to investigate the microstructure, phase composition and crystal orientation as well as material properties of Al-Mg alloyed additive. The results show that the grain size of Al-Mg alloyed additive becomes smaller with the decrease of welding current or increased welding speed. It is easier to obtain the additive parts with better grain uniformity with the increase of gas flow or interlayer residence time. The phase composition of Al-Mg alloyed additive consists of α-Al matrix and γ ($Al_{12}Mg_{17}$) phase. The eutectic reaction occurs during the additive manufacturing process, and the liquefying film is formed on the α-Al matrix and coated on the γ phase surface. The crystal grows preferentially along the <111> and <101> orientations. When the welding current is 90 A, the welding speed is 700 mm/min, the gas flow is 22.5 L/min and the interlayer residence time is 5 min, the Al-Mg alloy additive obtains the highest tensile strength. Under the optimal process parameters, the average grain size of Al-Mg alloyed additive is 25 μm, the transverse tensile strength reaches 382 MPa, the impact absorption energy is 26 J, and the corrosion current density is 3.485×10^{-6} A·cm^{-2}. Both tensile and impact fracture modes of Al-Mg alloyed additive are ductile fractures. From the current view, the Al-Mg alloys manufactured by WAAM-CMT have a better performance than those produced by the traditional casting process.

Keywords: wire arc additive manufacturing; cold metal transfer; Al-Mg alloys; orthogonal experiment; microstructure; mechanical properties

Citation: Liu, Y.; Liu, Z.; Zhou, G.; He, C.; Zhang, J. Microstructures and Properties of Al-Mg Alloys Manufactured by WAAM-CMT. Materials 2022, 15, 5460. https://doi.org/10.3390/ma15155460

Academic Editor: Emanuela Cerri

Received: 4 July 2022
Accepted: 6 August 2022
Published: 8 August 2022

Publisher's Note: MDPI stays neutral with regard to jurisdictional claims in published maps and institutional affiliations.

Copyright: © 2022 by the authors. Licensee MDPI, Basel, Switzerland. This article is an open access article distributed under the terms and conditions of the Creative Commons Attribution (CC BY) license (https://creativecommons.org/licenses/by/4.0/).

1. Introduction

Additive manufacturing (AM) technology is a bottom-up layer-by-layer manufacturing method based on digital models [1,2], which has been widely utilized in consumer electronics, automotive engineering, aerospace as well as other industries [3]. Wire and arc additive manufacturing (WAAM) is a kind of AM technology that utilizes arc as the heat source and metal wire as a stacking material to manufacture components [4]. There are numerous branches of AM technology, which are selected according to the performance of the additive metal, the shape and size of the additive product, etc. Selective laser sintering (SLS) [5] and selective laser melting (SLM) [6–9] are generally used for complex components with high forming accuracy and small volume. For metal components with high forming environment requirements and a high energy absorption rate, electron beam freeform fabrication (EBF^3) technology with wire feeding is generally adopted [10]. Compared with AM technologies such as SLS, SLM and EBF^3, WAAM has the technological advantages of low equipment energy consumption, high material utilization rate and high deposition efficiency. It is suitable for additive manufacturing of large-sized components and has been widely used by researchers at home and abroad [11,12].

Aluminum alloys have low density, high strength, good electrical conductivity, thermal conductivity and corrosion resistance as well as machinability, and are widely utilized in aerospace and transportation as well as civil industry [13]. Al-Mg alloys belong to the 5XXX series of aluminum alloys. Compared with other aluminum alloys, the Mg element, as a crucial strengthening element, improves the strength and the corrosion resistance of Al-Mg alloys [14,15]. However, traditional aluminum alloy forming methods such as casting, forging and machining face challenges in meeting the current demand for intricate structural components [16,17]. Therefore, the use of WAAM technology to manufacture aluminum alloy components has become the focus of urgent research. Cold metal transfer (CMT) is a new melt inert gas welding technology [18,19]. In the welding process, the droplet of the welding wire makes contact with the molten puddle, a short circuit occurs, and the arc voltage is instantaneously close to zero. During the short circuit transition, the CMT welding technology can pull the droplet through the wire pulling back to overcome the surface tension of the droplet and avoid droplet rejection and necking burst [20,21]. Compared with other welding technologies using arc as the heat source, CMT welding technology has the characteristics of low heat input and no spatter as well as good forming quality, and has received extensive attention in recent years [22–24]. Cong B [25] utilized WAAM-CMT technology to manufacture Al-6.3%Cu alloys to investigate the application potential of aluminum alloys. The results showed that pores were produced when CMT additive was used to manufacture aluminum alloys, but the porosity defects were effectively eliminated by setting appropriate welding process parameters. Gu J [26] utilized the interlayer rolling technology to process 5087 aluminum alloys produced by the WAAM-CMT technology, which significantly reduced the number of aluminum alloyed pores and provided a solution to the problem of porosity defects during the welding process of the aluminum alloys. Geng H [27] utilized WAAM technology to manufacture 5A06 aluminum alloys and found that the tensile properties of the additive parts were anisotropic, with a difference of 22 MPa between transverse and longitudinal tensile strength. Zhang C [28] used variable polarity CMT as an additive heat source of Al-6Mg alloys, and the grain size of Al-6Mg was 20.6–28.5 μm. In addition to investigating the defects and microstructures of aluminum alloys manufactured by WAAM-CMT, it is also of great significance to investigate additive properties. Horgar A [29] utilized WAAM-CMT technology and AA5183 Al-Mg alloys welding wire as filling material, and the tensile strength of the sample was 293 MPa. Gu J [30] manufactured 5087 Al-Mg alloyed components with the WAAM-CMT process, and the average microhardness of the alloys was 107.2 HV. Su C [31] utilized the WAAM-CMT process to manufacture the Al-Mg alloys and found that the crystal size on the surface of the alloys was 42.9–88.7 μm, while that on the inside of the alloys was 37.7–77.6 μm. The average tensile strength of the sample was about 255 MPa.

At present, some progress has been made in the research of the WAAM-CMT process for additive manufacturing of Al-Mg alloyed materials [32–34]. However, there are still gaps in the related study of the WAAM-CMT process for the additive manufacturing of 5556 Al-Mg alloys. To further expand the application of Al-Mg alloys in the field of additive manufacturing, a WAAM-CMT system was established to manufacture Al-Mg alloyed additive walls. Based on the orthogonal experiments, the process parameters of the welding current (WC), welding speed (WS), gas flow (GF) and interlayer residence time (IRT) were adjusted to investigate the walking path and optimal process parameters of Al-Mg alloyed additive. The evolution law of microstructures and the effect of different process parameters on the mechanical properties of Al-Mg alloyed additive were analyzed.

2. Materials and Methods

The WAAM-CMT system consists of a TranspulsSynergic 3200 CMT welder manufactured by Fronius Austria and a computer numerical control (CNC) system. The additive manufacturing experiment was carried out on a 5052 Al-Mg alloyed substrate with 200 mm × 180 mm × 6 mm using ER5556 Al-Mg alloyed wire with a diameter of 1.2 mm as filler wire. The chemical composition of welding wire and substrate is shown in Table 1. During the experiment, the welding torch was controlled by the CNC system to complete the X, Y and Z axis movement on the welding platform, and pure Ar was used as the experimental protective gas. The dry elongation of the welding wire was 12 mm, the walking distance along the X direction was set as 150 mm, and the increment in the Z direction (the distance of the welding torch elevation after each deposition) was set as 2 mm. The number of layers of the additive was 20. Before the experiment, the substrate surface was wiped with acetone and it was ensured that the welding wire was dry. Combined with the purpose of the investigation, WC, WS, GF and IRT were taken as the orthogonal experimental variables to optimize the parameters. The orthogonal experiment table of $L_{25}(5^4)$ was designed as shown in Table 2.

Table 1. Chemical composition of the substrate and filler material (mass fraction/%).

Materials	Si	Fe	Cu	Mn	Mg	Cr	Zn	Ti	Al
5052	0.25	0.40	0.10	0.10	2.2–2.8	0.15–0.35	0.10	0.01	Bal.
ER5556	0.07	0.17	0.01	0.61	4.90	0.11	0.08	0.09	Bal.

Table 2. Range of matrix building parameters with four factors and five levels.

Factors	Levels				
	1	2	3	4	5
WC/A	90	100	110	120	130
WS/mm·min^{-1}	400	500	600	700	800
GF/L·min^{-1}	12.5	15.0	17.5	20.0	22.5
IRT/min	1	2	3	4	5

The sampling location and size of the samples prepared for microstructure, tensile properties, and impact properties are shown in Figure 1. The microstructures of the Al-Mg alloyed wall (middle part) were observed. The sample was cut by a DK7763 super CNC wire-cutting machine according to the specified size, and the size of the metallographic piece was 10 mm × 2.5 mm × 1.5 mm. The size of the tensile specimen was determined according to GB/T 228.1-2021 "Metallic materials—Tensile testing—Part 1: Method of test at room temperature". After grinding, electrolytic polishing was carried out in the DF-3010 electrolytic polishing corrosion tester with corrosion voltage controlled at 20 V, and the corrosion reagent was an alcohol solution of 10% perchloric acid. Three groups of tensile specimens were cut along the parallel and perpendicular to the processing direction, and the tensile strength was measured and averaged. The impact sample size was determined according to GB/T 229-2020 "Metallic materials—Charpy pendulum impact test—Part 1: Test method". Three groups of impact samples were cut along the processing direction, and the impact absorption energy was measured and averaged.

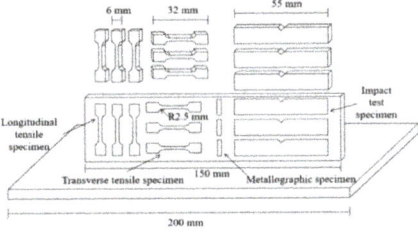

Figure 1. Schematic diagram of the sampling location and size.

A D/max-RB X-ray diffractometer (XRD) was utilized to analyze the phase composition of the additive. The scanning range was 5–90°. The microstructures of the Al-Mg alloyed components were analyzed by OLYMPUS-CK40M metallographic microscope (OM), S-4800 scanning electron microscope (SEM) and energy dispersive spectrometer (EDS). Electron back-scattered diffraction (EBSD) was used to obtain the orientation information of microscopic crystals in different process parameters and measure the average grain size. The testing equipment was S-4800 SEM with an EBSD probe. The data processing software of Channel 5 was used to calibrate the texture and crystal orientation of EBSD photos. At room temperature, the tensile test was carried out by a WDW-100B electronic universal testing machine at a tensile rate of 2 mm/min, and the impact test was carried out by the ZBC2602-C impact testing machine. The tensile and impact fracture morphologies were photographed by SEM. The PARSTAT-2273 electrochemical workstation was utilized to test the corrosion resistance of the sample by simulating seawater with a 3.5 wt% NaCl solution.

3. Results

3.1. Effect of the Walking Path on Al-Mg Alloyed Additive

The Al-Mg alloyed additive under different walking paths is shown in Figure 2. It is observed that the forming effect of the Al-Mg alloyed additive walls under the reciprocating walking path is better than that under the unidirectional walking path. The additive parts collapse at the end of each layer under the unidirectional walking path. The unidirectional additive is a discontinuous process. At the end of each welding layer, the welding torch is raised and returned to the starting position to continue depositing the next layer of metal. In additive deposition, the initial part of the additive goes through the arc burning and arc extinguishing process of the CMT welding machine successively so that Al-Mg alloyed welding wire is melted and deposited on the surface of the substrate.

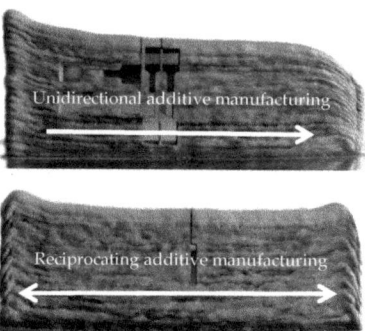

Figure 2. Al-Mg alloyed additive under different walking paths.

As the welding torch moves towards the specified deposition route, the arc burning and arc extinguishing processes are repeated, and the deposited Al-Mg alloys undergo the processes of heating and heat preservation as well as subcooling again. At the end of the additive, the metal is no longer subjected to an intricate thermal cycle as the welding torch discontinues moving forward because of arc extinction. Eventually, as the number of additive layers increases, collapse occurs at the end of unidirectional addition. It is observed that the WAAM walking path has a particular influence on the geometry of the additive. The continuous reciprocating way makes the thermal cycle at the beginning or end of the additive highly symmetrical but also improves the wetting effect during the additive process. Therefore, the Al-Mg alloyed additive manufactured by WAAM-CMT using the reciprocating walking path has a continuous forming, good appearance and no obvious welding defects, which avoids the problem of collapse in the additive process.

3.2. Optimal Process Parameters Based on Orthogonal Experiment

Based on the reciprocating additive manufacturing walking path, the transverse tensile strength (TTS) of Al-Mg alloyed additive components were measured under different process parameters in the orthogonal experiment. The results are shown in Table 3. The direction of transverse tensile strength is the walking direction of the welding torch. The effect of thermal cycling conditions on the alloys is the same in this direction, so the microstructural distribution of the alloys is more uniform. The TTS is finally taken as the criterion to evaluate the mechanical properties of the orthogonal experiments.

Table 3. Experimental results of transverse tensile strength under orthogonal process parameters.

No.	WC/A	WS/mm·min^{-1}	GF/L·min^{-1}	IRT/min	TTS/MPa	No.	WC/A	WS/mm·min^{-1}	GF/L·min^{-1}	IRT/min	TTS/MPa
1	90	400	12.5	1	331	17	120	500	22.5	3	296
2	90	500	15.0	2	316	18	120	600	12.5	4	284
3	90	600	17.5	3	304	19	120	700	15.0	5	319
4	90	700	20.0	4	337	20	120	800	17.5	1	311
5	90	800	22.5	5	367	21	130	400	22.5	4	307
6	100	400	15.0	3	361	22	130	500	12.5	5	308
7	100	500	17.5	4	337	23	130	600	15.0	1	285
8	100	600	20.0	5	309	24	130	700	17.5	2	329
9	100	700	22.5	1	323	25	130	800	20.0	3	293
10	100	800	12.5	2	300						
11	110	400	17.5	5	309	k_1	331.24	316.37	305.60	310.30	
12	110	500	20.0	1	315	k_2	326.15	314.40	310.56	310.56	
13	110	600	22.5	2	311	k_3	309.49	298.63	299.57	299.57	
14	110	700	12.5	3	304	k_4	296.71	322.48	305.68	319.57	
15	110	800	15.0	4	309	k_5	304.27	315.98	320.60	327.86	
16	120	400	20.0	2	274	R	34.53	23.85	21.03	28.29	

K_i is the average value of all TTS results of this factor at the level of i, and R is the range of K_i. In other words, $R = k_{max} - k_{min}$. The larger k_i indicates the optimal experimental results at the i level. The larger R is, the greater the influence of this factor on the experimental results is. According to the range analysis results, the optimal process parameters at the highest TTS are as follows: WC = 90 A, WS = 700 mm/min, GF = 22.5 L/min, IRT = 5 min. At this time, the TTS of Al-Mg alloyed additive can reach 382 MPa. Under the experimental parameters, WC is the most crucial factor affecting the TTS of Al-Mg alloyed additive.

4. Discussion

4.1. Pore Analysis of Al-Mg Alloyed Additive under Orthogonal Experiment

The metallographic microstructures and porosity of Al-Mg alloyed walls are shown in Table 4. The black granular material is a porosity defect formed during the additive process. Due to the good thermal conductivity of aluminum alloys, the molten puddle cooling speed is extremely fast. The escape time of bubbles in the molten puddle is not sufficiently fast, and the molten puddle begins to crystallize and solidify, thus forming the porosity defects. According to Li Z et al. [35], the pores formed in the aluminum alloyed additive process are especially hydrogen pores. The hydrogen pores mainly generate heterogeneous nucleated particles at grain boundaries and grow up through free diffusion and merger. The results demonstrate a competitive growth relationship between the formation of pores and the crystal structure. When the heat input of the additive is smaller, the grain size of the alloys becomes more minor, and the size and number of pores generated are smaller. On the contrary, the larger the additive heat input, the larger the number and size of pores.

Table 4. Al-Mg metallographic structure under orthogonal process parameters.

Al-Mg Metallographic Structure and Porosity under Orthogonal Experiment [1] 200 Mm				
01# 0.7%	02# 1.6%	03# 1.0%	04# 0.7%	05# 0.5%
06# 1.9%	07# 1.9%	08# 0.9%	09# 0.8%	10# 1.3%
11# 2.0%	12# 1.2%	13# 0.8%	14# 2.0%	15# 1.7%
16# 5.4%	17# 1.8%	18# 1.3%	19# 1.1%	20# 1.1%
21# 2.0%	22# 0.5%	23# 0.4%	24# 0.8%	25# 3.0%

[1] The specific values of 01#–25# orthogonal process parameters correspond to Table 3 above.

It can be seen from Table 4 that the formation of porosity is closely related to the heat input during the additive process. Porosity is the total area of the pore divided by that of metallography. To ensure the accuracy of the experiment, the average value of each parameter sample was taken from three groups. To improve the tensile strength of the alloy, the porosity must be reduced. The porosity is the highest under the experimental parameters of 16# (WC = 120 A, WS = 400 mm/min, GF = 20 L/min, IRT = 5 min), and the higher porosity results in a sharp decrease in the TTS under 16# orthogonal parameters. At the same time, it is observed that under the practical parameters of 5# (WC = 90 A, WS = 800 mm/min, GF = 22.5 L/min, IRT = 5 min), the porosity is the lower, and the corresponding TTS is the highest. The reason is that under the experimental parameters of 5#, the welding current is small, and the welding speed is considerable, resulting in low heat input in the depositing process. On the one hand, the lower the welding heat input, the less the elements evaporated due to overheating during the additive process, and the fewer bubbles were generated in the molten puddle, which is the main factor in avoiding the formation of pores. On the other hand, the smaller the welding heat input, the shorter the residence time of the molten puddle at a high temperature, and the more sufficient the escape time of tiny bubbles in the molten puddle was, to reduce the porosity and give the Al-Mg alloyed additive better mechanical properties. In addition, compared with the experimental parameters of 16#, the experimental parameters of 5# have a faster gas flow protection to isolate the gas in the outside air from penetrating the Al-Mg alloys and strengthen the security of the alloyed structure. Therefore, under the WAAM-CMT process, decreasing the welding current or increasing the welding speed is to decrease the heat input. WAAM-CMT technology can effectively reduce or even eliminate pore defects due to the reduction of heat input, thus enhancing the mechanical properties of Al-Mg alloyed additives.

4.2. Grain Size Analysis of Al-Mg Alloyed Additive under Different Process Parameters

The influence of different process parameters on the EBSD microstructure of Al-Mg alloyed additive was studied by single-factor analysis, and the test results are shown in Figure 3. The grains with different grain orientations were marked with different colors by EBSD to analyze the grain sizes. Figure 3a shows the EBSD microstructure of Al-Mg alloyed additive manufactured under the process parameters of WC = 90 A, WS = 800 mm/min, GF = 22.5 L/min as well as IRT = 5 min. Taking Figure 3a as the contrasting sample, the process conditions in Figure 3b–e only change the welding current, welding speed and gas flow, as well as interlayer residence time, while keeping other process parameters unchanged.

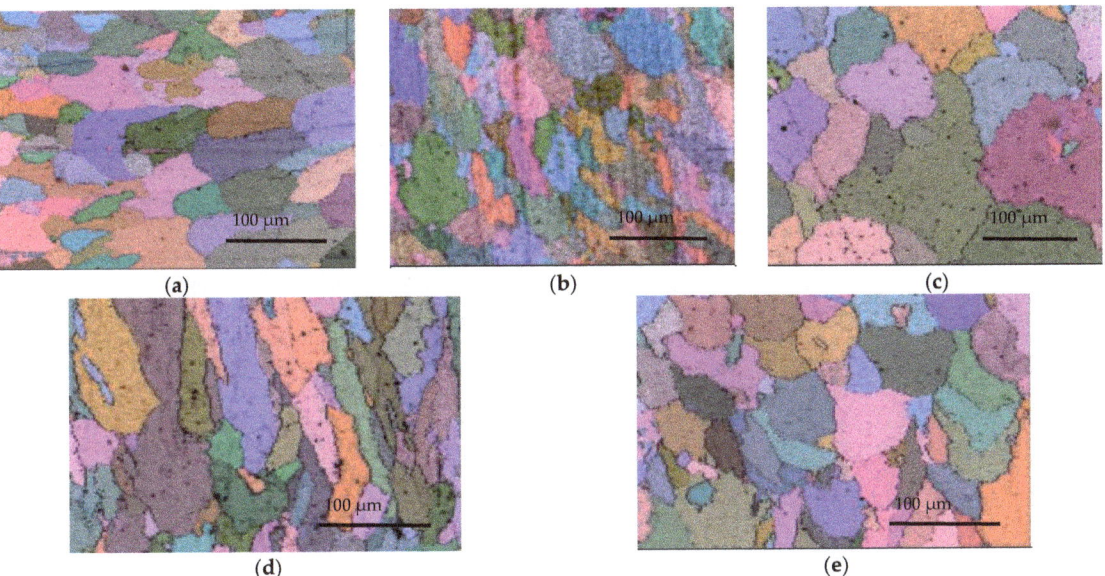

Figure 3. EBSD microstructure of Al-Mg alloyed additive under different process parameters: (a) Contrasting sample (90 A, 800 mm/min, 22.5 L/min, 5 min); (b) 130 A welding current; (c) 400 mm/min welding speed; (d) 12.5 L/min gas flow; (e) 1 min interlayer residence time.

Figure 3b shows the EBSD microstructure of the additive under the welding current of 130 A. Compared with Figure 3a (90 A), the microstructures of the additive become coarser with the increase of welding current. The reason is that the growth of welding current increases the heat input in the WAAM-CMT process. The rise in the heat input increases the heat dissipation time of the additive, and the grain has sufficient time to grow. Therefore, the grain size is relatively coarse when the welding current is 130 A. Figure 3c shows the EBSD microstructure of additive at a welding speed of 400 mm/min. In the WAAM-CMT process, welding speed is also one of the crucial factors affecting the heat input of additive. Compared with Figure 3a (800 mm/min), when the welding speed is slower, the volume of arc melting wire per unit length is more extensive, resulting in increased additive heat input and grain coarsening of the Al-Mg alloyed additive. Therefore, decreasing welding current or increasing welding speed can reduce the heat input of the welding process, refine grain size and form an equiaxed grain structure.

Figure 3d shows the EBSD microstructure of the additive at a gas flow rate of 12.5 L/min. Compared with Figure 3a (22.5 L/min), the grains are columnar, and the size distribution is uneven when the gas flow rate is slow. The reason is that the shielding gas plays a role in isolating the outside air and can also play a role in optimizing the weld bead. When the gas flow is languid, the weld bead is weakened by the gas pressure, causing the grains to grow outward in the form of columnar crystals. Figure 3e shows the EBSD microstructure of the additive under the interlayer residence time of 1 min. Compared with Figure 3a (5 min), it is found that the microstructural microstructure obtained under different interlayer residence time is good. Still, when the interlayer residence time is 5 min, the grain size distribution of the additive is more uniform. The reason is that interlayer residence time plays an essential role in improving the thermal accumulation effect of additive parts. The shorter the residence time between layers, the less heat accumulated in each layer is lost. The accumulated heat will prolong the holding time of grains, and the grains of Al-Mg alloyed additive have sufficient conditions to grow. As the number of additive layers increases, the heat accumulation effect becomes more evident. Therefore, increasing gas flow rate and prolonging interlayer residence time is beneficial for obtaining the grains with good uniformity.

In the WAAM-CMT process, the additive heat input is the fundamental factor affecting the crystal size of the additive, and the additive heat input is mainly affected by the welding current and welding speed. As the welding current decreases from 130 A to 90 A, or as the welding speed increases from 400 mm/min to 800 mm/min, the grain size of Al-Mg alloys becomes smaller. The results of the EBSD grain size analysis demonstrate that the microstructures of Al-Mg alloyed additive are mainly equiaxed grain and columnar grain. Due to the existence of columnar crystals, the Al-Mg alloyed additive is anisotropic. As the gas flow rate increases from 12.5 L/min to 22.5 L/min, the equiaxed grains are easily obtained. The reduction of columnar crystals gives the material better microstructures and properties. With the interlayer residence time increasing from 1 min to 5 min, it is easier to obtain Al-Mg alloyed additive with good microstructural uniformity and fine grain size. The results of orthogonal tensile experiments demonstrate that the optimal tensile strength of Al-Mg alloys is brought under the conditions of 90 A welding current, 700 mm/min welding speed, 22.5 L/min gas flow rate and 5 min interlayer residence time. The average grain size of the additive is 25 μm under the optimal process parameters measured by the transverse scribing method in Channel 5 software, and the measurement method is shown in Figure 4.

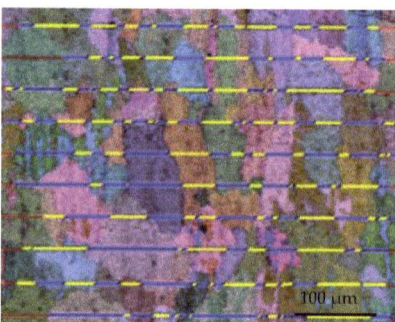

Figure 4. Grain size of Al-Mg alloyed additive measured by scribing method.

4.3. Phase Analysis of Al-Mg Alloyed Additive

The XRD pattern of the Al-Mg alloyed additive is shown in Figure 5. It can be seen that there are only diffraction peaks of the α-Al phase and a small amount of $Al_{12}Mg_{17}$

phase in the XRD pattern of the sample. Due to the relatively small content of other elements, they do not appear in the XRD pattern. Under the action of the CMT welding heat source, Al-Mg alloyed welding wire melts to form a molten puddle. As the CMT welding technology has the characteristics of short-circuit transition, the decrease of arc and droplet temperature leads to the extremely short existence of the molten puddle. The molten puddle, after supercooling, will crystallize and precipitate on the substrate or the previous Al-Mg alloyed layer. When the temperature reaches between eutectic temperature and liquidus temperature, the material will undergo the eutectic reaction: $L \rightleftharpoons Mg + \gamma\ (Al_{12}Mg_{17})$.

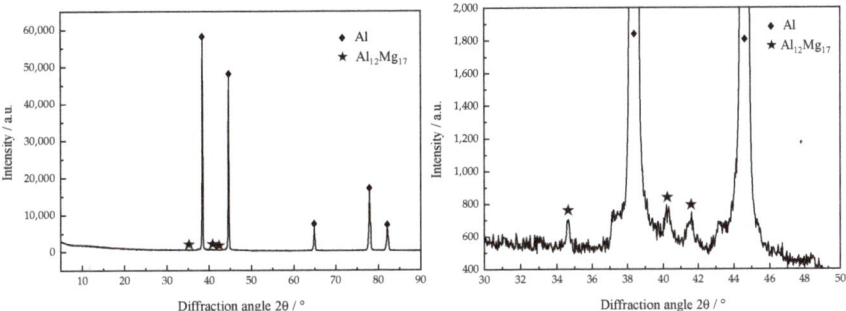

Figure 5. XRD pattern of Al-Mg alloyed additive.

Although the heat accumulated by the last metal layer on the previous metal layer is increasing, the area of the molten puddle cooling to the surrounding environment is also increasing. When the number of layers increases to a certain height, the sediments stay in thermal equilibrium. The heat accumulation in the WAAM-CMT process is the main reason for the microstructural transformation. Under the heat accumulation effect, the preheating effect of the former layer of metal on the last layer of metal is gradually strengthened, which leads to the extension of the holding time for the growth of various microstructures in Al-Mg alloys, thus forming larger grains. With the increase of additive layers, the heat absorption and dissipation of the molten puddle reaches a balanced state in unit time, and the heat accumulation effect gradually tends to be gentle. Therefore, the interception position of the metallographic sample is the 10th layer of the additive wall.

Line energy is an essential means to characterize heat input. The calculation formula of line energy is $q = UI/V$, where U is the welding voltage, I is the welding current, and V is the welding speed. Under WC = 90 A, GF = 20 L/min as well as IRT = 2 min, the SEM microstructure of Al-Mg alloyed additive with different line energy is shown in Figure 6.

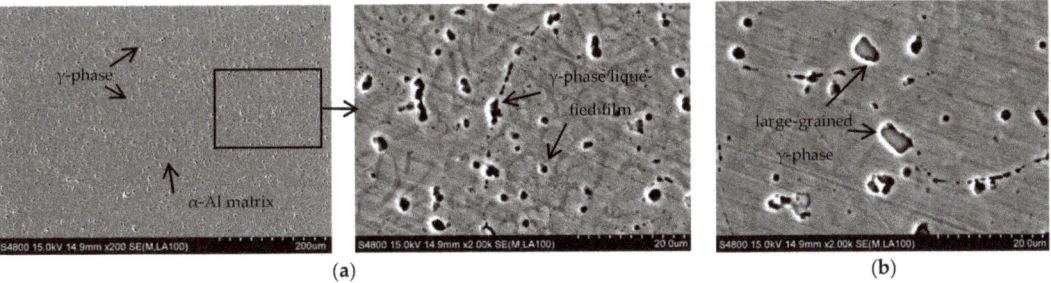

Figure 6. SEM microstructure of Al-Mg alloyed additive under different linear energies: (**a**) 1440 J/cm line energy; (**b**) 1800 J/cm line energy.

Figure 6a shows the SEM microstructure of the additive at different magnifications under the condition of 1440 J/cm line energy. It is observed that the α-Al phase is the alloyed matrix, and there is a region of liquefaction. In addition, the eutectic reaction occurs to form the Mg and γ phase. When the next layer of metal is deposited, the upper layer of metal is liquefied again by the Mg and γ phase due to heating then redissolved into the molten α-Al matrix. However, due to the extremely short time of the CMT welding heat source, the solid-state phase transformation of the γ phase is minimal, so the γ phase does not have enough time to dissolve into the α-Al matrix fully. When the temperature is reduced to the eutectic reaction temperature, the Mg and γ phase will crystallize again, forming the eutectic liquid phase. As the eutectic reaction time is extremely short, the liquefied film is formed at the part of the original Mg and γ phase as the temperature continues to decrease. Figure 6b shows the SEM microstructure of the additive under the condition of 1800 J/cm line energy. Compared with 1440 J/cm line energy, with the increase of heat input, the nucleation of γ phase crystal is not uniform, accompanied by the appearance of a large-grained γ phase microstructure. To determine the composition of black particles in the SEM microstructure of Al-Mg alloyed additive, EDS point scanning was performed for the black particles, as shown in Figure 7. EDS point scanning results demonstrate that Al and Mg are the main elements in the particles. Combined with the XRD pattern and analysis results, it is verified that the black substance is the γ phase microstructure. The microstructures are improved by reducing the heat input, thereby increasing the mechanical properties of the Al-Mg alloyed additive. On the one hand, reducing the heat input can avoid the formation of the γ phase with large grains. On the other hand, the decrease of heat input makes the grains fine and uniform, and the dislocation movement is hindered by the effect of fine-grain strengthening to improve the mechanical properties of Al-Mg alloyed additive.

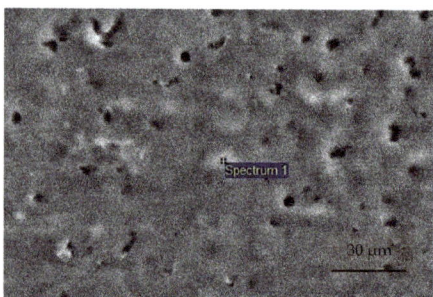 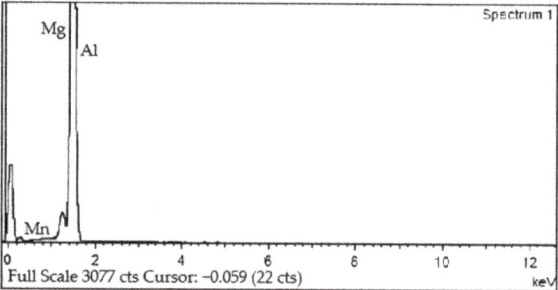

Figure 7. EDS point scanning results of Al-Mg alloyed additive.

The chemical composition of the additive is shown in Table 5. The results show that the Mg element content decreases from 4.9% to 4.68%, and the loss rate is 4.49%. Yuan, T [36] fabricated the Al-Mg alloy by the WAAM-TIG technology and studied the loss of the Mg element under different process parameters. The results show that the loss rate of the Mg element is 5.56% at the optimum process parameters. Because the WAAM-CMT has the significant advantage of the short circuit transition, the loss of Mg element in the Al-Mg alloy is slightly reduced compared with the WAAM-TIG.

Table 5. Chemical composition of the substrate and additive material (mass fraction/%).

Materials	Si	Fe	Cu	Mn	Mg	Cr	Zn	Ti	Al
5052	0.25	0.39	0.11	0.10	2.54	0.21	0.09	0.01	Bal.
5556	0.08	0.18	0.01	0.63	4.68	0.12	0.07	0.10	Bal.

4.4. Crystal Orientation Analysis of Al-Mg Alloyed Additive

Using EBSD crystal orientation imaging technology to obtain crystal orientation measurement data is beneficial for analyzing the change of texture in the sample. Figure 8 is the polar diagram of Al-Mg alloyed additive on different planes when the TTS is optimal. It can be seen that there is a <111> texture on the XY plane at 45° off the X direction, while there is no preferred orientation on other crystal planes. The results indicate that Al-Mg alloyed additive grows preferentially in the <111> orientation, which is consistent with the close-packed plane of α-Al with the face-centered cubic crystal structure.

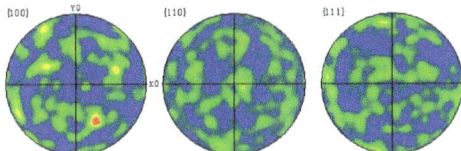

Figure 8. Polar diagrams of Al-Mg alloyed additive on different planes.

To further verify the rationality of the analysis, the samples were processed with an inverse pole diagram as shown in Figure 9. By calibrating the crystal coordinates with the samples, it is found that the texture of the sample exists not only in the <111> orientation but also in the <101> orientation. However, the crystal orientation intensity of the <101> is less than that of the <111>. The <101> plane is the close-packed plane of the γ phase with the body-centered cubic structure, indicating that the crystal grows preferentially along the <111> plane during the additive process. On the contrary, if the γ phase is preferentially extended, the liquefied film will not form at the part of the Mg and γ phase. Therefore, in the process of manufacturing Al-Mg alloyed additive with WAAM-CMT technology, the crystals grow preferentially along the <111> and <101> orientations. In contrast, the crystals which deviate significantly from the <111> and <101> orientations will discontinue growing. The crystal orientation distribution of the Al-Mg alloyed additive is shown in Figure 10. In the aluminum alloyed matrix, Mg mainly grows along with the <101> orientation and forms the γ phase, resulting in lattice distortion, which significantly strengthens the hinderance of dislocation movement and has a solid solution strengthening effect on the alloys.

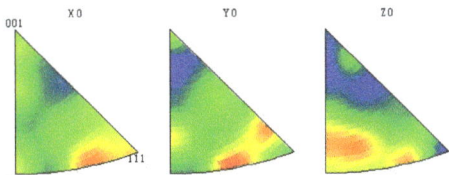

Figure 9. Inverse pole diagrams of Al-Mg alloyed additive on different planes.

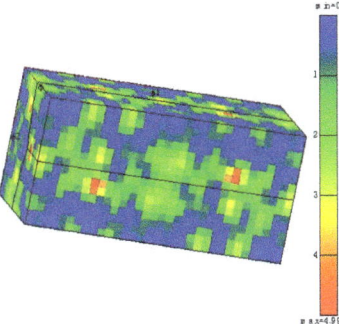

Figure 10. Orientation distribution diagram of Al-Mg alloyed additive.

4.5. Performance Analysis of Al-Mg Alloyed Additive

4.5.1. Analysis of Tensile Properties and Fracture Morphology

The transverse and longitudinal tensile strengths of the Al-Mg alloyed additive measured under different process parameters of the orthogonal experiment in Table 3 above are shown in Figure 11.

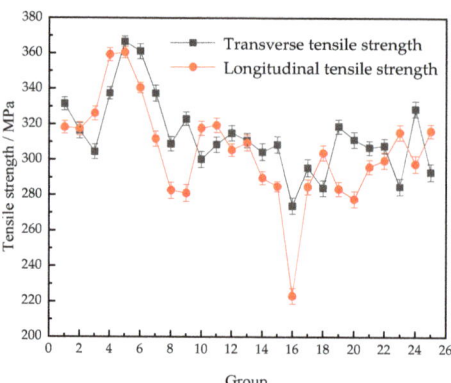

Figure 11. Tensile strength of Al-Mg alloyed additive under different process parameters.

There is little difference between the transverse and longitudinal tensile strengths of the Al-Mg alloyed additive on the whole. Still, the transverse tensile strength of the additive is slightly higher than the longitudinal tensile strength. The reason is that in the process of the additive, each layer of metal in the transverse direction is heated to melt and supercooled to crystallize, forming a uniform and stable Al-Mg alloyed microstructure. In the longitudinal order, the problem of interlayer microstructural combination needs to be considered. Due to the longitudinal gradient diffusion of heat centered on the welding torch and the gravity effect of liquid metal, the heating cycle and heat accumulation conditions of the lower metal and the upper metal are different, so the longitudinal interlayer microstructure combination is worse than the transverse interlayer microstructure combination.

The microstructures of the Al-Mg alloyed additives manufactured by WAAM-CMT technology are anisotropic, but the effect is not obvious. The average tensile strength of the Al-Mg alloyed additives is 310 MPa, which is approximately 30% higher than that of the matrix. Traditional Al-Mg alloys are manufactured by stabilizing annealing or work hardening process, and the tensile strength is generally between 200 MPa and 400 MPa [37]. The optimum tensile strength of the Al-Mg alloyed additive manufactured by WAAM-CMT technology is 382 MPa, which is higher than that of the annealed Al-Mg alloys and some machined Al-Mg alloyed forgings. The tensile stress–strain curve of the optimal parameters is shown in Figure 12. The experimental results can provide the experimental basis and data reference for the manufacturing process of Al-Mg alloys by WAAM-CMT.

The SEM morphology of transverse tensile fracture of the additive is shown in Figure 13. It can be seen that the cross-sectional area at the rupture of the tensile specimen shrinks, and the necking phenomenon occurs. It demonstrates that during resisting tensile deformation, dislocations are constantly produced inside the material, and the dislocations at grain boundaries or γ phase particles are used to fight the tensile deformation caused by the outside of the material. The fracture surface of the sample is gray–white, with a cup-cone shape around it and a tearing edge. A large number of dimples were observed in the fracture under a high-power microscope. Therefore, the fracture mode of the Al-Mg alloyed additive is a ductile fracture. In general, the formation of the dimple of Al-Mg alloys is related to the precipitation of the γ phase, and the size, microstructure and distribution of the γ phase affect the fracture mode of the material.

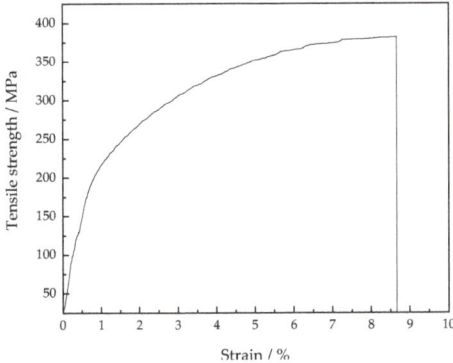

Figure 12. The tensile stress-strain curve of the optimal parameters.

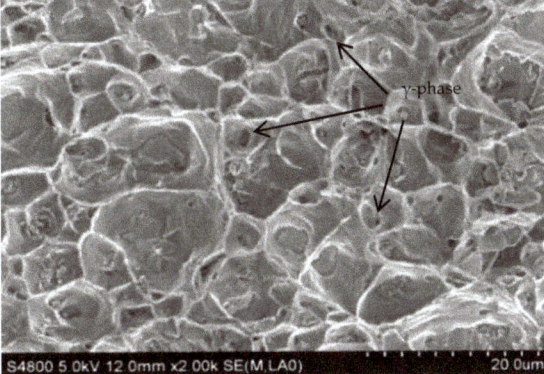

Figure 13. SEM morphology of transverse tensile fracture of Al-Mg alloyed additive.

4.5.2. Analysis of Impact Properties and Fracture Morphology

The impact property is also a crucial mechanical property index for Al-Mg alloyed additive according to the different service environments of materials. Under WC = 90 A, WS = 700 mm/min, GF = 22.5 L/min and IRT = 5 min, the average impact absorption energy of Al-Mg alloyed additive measured at room temperature is 26.19 J, which is higher than the standard value of 24.00 J of Al-Mg alloys. Yan X [38] adopted the plasma arc welding process to weld the 5052 Al-Mg alloyed substrates with ER5356 Al-Mg alloyed welding wire, and the average impact absorption energy was 19.60 J. Compared with this process, the impact property of Al-Mg alloyed additive manufactured by WAAM-CMT technology is improved by 22%.

The impact fracture morphology of the additive is shown in Figure 14. There are many dimples at the impact fracture, which indicates that the fracture mode is ductile fracture. The Al-Mg alloy grains manufactured by WAAM-CMT are mainly equiaxed. Once subjected to the impact load, the grains will produce plastic deformation along the shear slip direction. Compared with the SEM microstructure of the Al-Mg alloyed additive in Figure 6 above, the grains are elongated to produce the small-angle grain boundaries under the action of shear stress. The large-angle grain boundaries are generated from the small-angle grain boundaries with plastic deformation, which is finally manifested as the microscopic morphology in Figure 14. The sediments formed at grain boundaries become the core of grain recrystallization and promote the impact deformation resistance of microstructures. Under the impact load, the dislocation cannot resist the impact load, and fracture failure occurs at the weak position, such as the grain boundary, before it has sufficient time to accumulate to the grain boundary or γ phase fully. In conclusion, Al-Mg alloyed additive manufactured by WAAM-CMT technology has excellent mechanical properties and can meet the basic requirements of modern production and manufacturing.

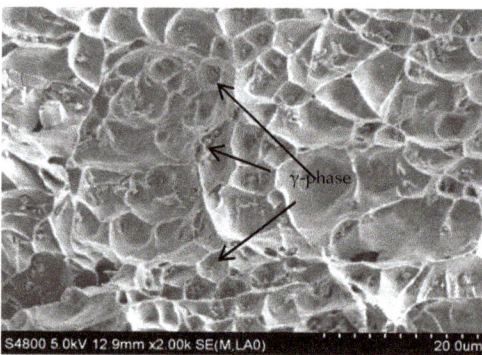

Figure 14. SEM morphology of impact fracture of Al-Mg alloyed additive.

4.5.3. Corrosion Resistance Analysis

Electrochemical tests were carried out on 5556 Al-Mg alloyed additive and 5052 Al-Mg alloyed experimental substrates, and the polarization curves were measured as shown in Figure 15. It is observed that the corrosion current density (I_{corr}) of Al-Mg alloyed additive is 3.485×10^{-6} A·cm^{-2}, while that of Al-Mg experimental substrate is 4.685×10^{-6} A·cm^{-2}. The corrosion current density of Al-Mg experimental substrate is slightly higher than that of the additive. The higher the corrosion current density, the faster the corrosion rate of the material. In addition, the corrosion potential (E_{corr}) of the 5556 additive is -0.95 V, while that of the substrate is -0.99 V. The more positive the corrosion potential, the stronger the corrosion inertness of the material. Therefore, the additive provides better corrosion resistance than the substrate. Compared with the α-Al matrix, the γ phase is the anode, and the corrosion reaction occurs preferentially. Therefore, the alloys have a great sensitivity to intergranular corrosion.

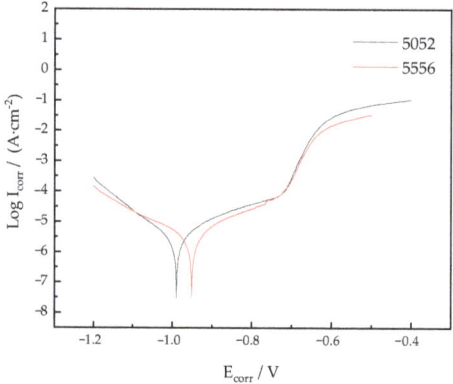

Figure 15. Polarization curves of Al-Mg alloys.

References [39,40] show that the properties of the Al-Mg alloy are improved with the increase of the Mg element content. However, when the content of the Mg element exceeds 7%, the precipitates of the Mg element increase and the crystallization hot crack occurs. In the study, the Mg content of the 5556 Al-Mg additive manufactured by WAAM-CMT is

4.68%, while the Mg content of the 5052 Al-Mg substrate is 2.54%. Since the Mg content of the 5556 Al-Mg additive is higher than that of the 5052 Al-Mg alloy, the γ phase with a relatively large volume fraction acts as a corrosion barrier to inhibit the corrosion of the matrix. Therefore, the 5556 Al-Mg alloy additive has better corrosion resistance. From the current view, the 5052 Al-Mg alloy is one of the most commonly used Al-Mg alloys. Because of its good corrosion resistance, it is widely utilized in the manufacturing of aircraft fuel tanks and ship sheet metal components. According to the polarization curve data, the corrosion resistance of 5556 Al-Mg alloyed additive is 1.34 times that of 5052 Al-Mg alloyed substrate. Therefore, 5556 Al-Mg alloyed additives can be operated in a particular seawater corrosion environment.

5. Conclusions

(1) For Al-Mg alloyed additive manufactured by WAAM-CMT, the forming effect of the reciprocating walking path is better than that of the unidirectional walking path. Based on the orthogonal experiment, the optimal process parameters for the highest tensile strength are as follows: 90 A of welding current, 700 mm/min of welding speed, 22.5 L/min of gas flow and 5 min of interlayer residence time;

(2) As the welding current decreases from 130 A to 90 A, the welding speed increases from 400 mm/min to 800 mm/min, the gas flow increases from 12.5 L/min to 22.5 L/min, the interlayer residence time increases from 1 min to 5 min, and the grain size of Al-Mg alloyed additive is gradually refined. The microstructures are mainly equiaxed and columnar crystals. Under the optimal process parameters, the average grain size of Al-Mg alloyed additive is 25 µm;

(3) Al-Mg alloyed additive is mainly composed of the α-Al phase and a small amount of $Al_{12}Mg_{17}$ phase, and the eutectic reaction will occur in the additive process: $L \rightleftharpoons Mg + \gamma$ ($Al_{12}Mg_{17}$). Mg exists in the form of $Al_{12}Mg_{17}$ solid solution in Al-Mg alloyed additive. With the alternation of the thermal cycle, the eutectic reaction results in liquefied film at the original Mg and γ phase locations. The larger the welding heat input, the larger the grain size of the liquid film. During the process of WAAM-CMT manufacturing, the crystals of Al-Mg alloyed additive mainly grow preferentially along the orientations of <111> and <101>. At the same time, the preferred strength of <111> is greater than that of <101>;

(4) The maximum tensile strength of Al-Mg alloyed additive is 382 MPa, the average tensile strength is 310 MPa, and the average impact absorption energy is 26.19 J. Under static tensile load or impact load, ductile fracture occurs in all Al-Mg alloyed additive parts, and there are many dimples in the fracture. The corrosion resistance of the Al-Mg alloyed additive is better than that of the substrate.

Author Contributions: Conceptualization, Y.L. and Z.L.; methodology, Y.L., Z.L. and G.Z.; software, Z.L. and G.Z.; validation, Y.L., C.H. and J.Z.; formal analysis, Z.L. and G.Z.; investigation, Y.L., Z.L. and J.Z.; resources, Y.L., C.H. and J.Z.; data curation, Z.L. and G.Z.; writing—original draft preparation, Y.L. and G.Z.; writing—review and editing, Y.L., Z.L. and G.Z.; visualization, C.H. and J.Z.; supervision, Y.L., C.H. and J.Z.; project administration, Y.L.; funding acquisition, Y.L. and C.H. All authors have read and agreed to the published version of the manuscript.

Funding: This study was financially funded by the Support Program for Innovative Talents in Universities of Liaoning Province (Grant number: LR2019042), and the Key R & D Project of Liaoning Province (Grant number: 2020JH2/10100011).

Institutional Review Board Statement: Not applicable.

Informed Consent Statement: Not applicable.

Data Availability Statement: Not applicable.

Acknowledgments: We sincerely appreciate the support provided by Shenyang University in the preparation of this manuscript.

Conflicts of Interest: The authors declare no conflict of interest.

References

1. Zhu, J.; Zhou, H.; Wang, C.; Zhou, L.; Yuan, S.; Zhang, W. A review of topology optimization for additive manufacturing: Status and challenges. *Acta Aeronaut. Astronaut. Sin.* **2021**, *34*, 91–110. [CrossRef]
2. Kok, Y.; Tan, X.P.; Wang, P.; Nai, M.L.S.; Loh, N.H.; Liu, E.; Tor, S.B. Anisotropy and heterogeneity of microstructure and mechanical properties in metal additive manufacturing: A critical review. *Mater. Des.* **2017**, *139*, 565–586. [CrossRef]
3. Tuan, D.; Alireza, K.; Gabriele, I.; Kate, T.Q.N.; David, H. Additive manufacturing (3D printing): A review of materials, methods, applications and challenges. *Compos. Part B Eng.* **2018**, *143*, 172–196.
4. Derekar, K.S. A review of wire arc additive manufacturing and advances in wire arc additive manufacturing of aluminium. *Mater. Sci. Technol.* **2018**, *34*, 895–916. [CrossRef]
5. Chen, A.N.; Wu, J.M.; Liu, K.; Chen, J.Y.; Xiao, H.; Chen, P.; Li, C.H.; Shi, Y.S. High-performance ceramic parts with complex shape prepared by selective laser sintering: A review. *Adv. Appl. Ceram.* **2018**, *117*, 100–117. [CrossRef]
6. Zhang, J.; Song, B.; Wei, Q.; Bourell, D.; Shi, Y. A review of selective laser melting of aluminum alloys: Processing, microstructure, property and developing trends. *J. Mater. Sci. Technol.* **2019**, *35*, 270–284. [CrossRef]
7. Li, R.D.; Chen, H.; Zhu, H.B.; Wang, M.B.; Chen, C.; Yuan, T.C. Effect of aging treatment on the microstructure and mechanical properties of Al-3.02Mg-0.2Sc-0.1Zr alloy printed by selective laser melting. *Mater. Des.* **2019**, *168*, 107668. [CrossRef]
8. Churyumov, A.Y.; Pozdniakov, A.V.; Prosviryakov, A.S.; Loginova, I.S.; Daubarayte, D.K.; Ryabov, D.K.; Korolev, V.A.; Solonin, A.N.; Pavlov, M.D.; Valchuk, S.V. Microstructure and mechanical properties of a novel selective laser melted Al–Mg alloy with low Sc content. *Mater. Res. Express* **2019**, *6*, 126595. [CrossRef]
9. Shen, X.F.; Cheng, Z.Y.; Wang, C.G.; Wu, H.F.; Yang, Q.; Wang, G.W.; Huang, S.K. Effect of heat treatments on the microstructure and mechanical properties of Al-Mg-Sc-Zr alloy fabricated by selective laser melting. *Opt. Laser Technol.* **2021**, *143*, 107312. [CrossRef]
10. Li, S.W.; Gao, Q.W.; Zhao, J.; Liu, H.W.; Wang, P.F. Research progress and prospect of electron beam freeform fabrication. *Mater. China* **2021**, *40*, 130–138.
11. Herawan, S.G.; Rosli, N.A.; Alkahari, M.R.; Abdollah, M.F.B.; Ramli, F.R. Review on effect of heat input for wire arc additive manufacturing process. *J. Mater. Res. Technol.* **2021**, *11*, 2127–2145.
12. Cunningham, C.R.; Flynn, J.M.; Shokrani, A.; Dhokia, V.; Newman, S.T. Invited review article: Strategies and processes for high quality wire arc additive manufacturing. *Addit. Manuf.* **2018**, *22*, 672–686. [CrossRef]
13. Tawfik, M.M.; Nemat-Alla, M.M.; Dewidar, M.M. Enhancing the properties of aluminum alloys fabricated using wire + arc additive manufacturing technique—A review. *J. Mater. Res. Technol.* **2021**, *13*, 754–768. [CrossRef]
14. Su, D.; Zhang, J.; Wang, B. The microstructure and weldability in welded joints for AA 5356 aluminum alloy after adding modified trace amounts of Sc and Zr. *J. Manuf. Process.* **2020**, *57*, 488–498. [CrossRef]
15. Gaur, V.; Enoki, M.; Okada, T.; Yomogida, S. A study on fatigue behavior of MIG-welded Al-Mg alloy with different filler-wire materials under mean stress. *Int. J. Fatigue* **2018**, *107*, 119–129. [CrossRef]
16. Chen, Y.; Wang, H.; Wang, X.; Ding, H.; Zhao, J.; Zhang, F.; Ren, Z. Influence of tool pin eccentricity on microstructural evolution and mechanical properties of friction stir processed Al-5052 alloy. *Mater. Sci. Eng.* **2019**, *739*, 272–276. [CrossRef]
17. Majeed, T.; Mehta, Y.; Siddiquee, A.N. Precipitation-dependent corrosion analysis of heat treatable aluminum alloys via friction stir welding, a review. *Proc. Inst. Mech. Eng. Part C J. Mech. Eng. Sci.* **2021**, *235*, 7600–7626. [CrossRef]
18. Mari, D.; Oli, T.; Kondi, V.; Samardi, I. Statistical analysis of MAG-CMT welding parameters and their influence on the Ni-alloy weld overlay quality on 16Mo3 base material. *Weld. World* **2022**, *66*, 815–831. [CrossRef]
19. Chen, X.; Su, C.; Wang, Y.; Arshad, N.S.; Konovalov, S.; Jayalakshmi, R.; Arvind, S. Cold Metal Transfer (CMT) Based Wire and Arc Additive Manufacture (WAAM) System. *J. Surf. Investig. X Ray Synchrotron Neutron Tech.* **2018**, *12*, 1278–1284. [CrossRef]
20. Han, S.; Zhang, Z.; Ruan, P.; Cheng, S.; Xue, D. Fabrication of circular cooling channels by cold metal transfer based wire and arc additive manufacturing. *Proc. Inst. Mech. Eng. Part B J. Eng. Manuf.* **2021**, *235*, 1715–1726. [CrossRef]
21. Tian, Y.; Shen, J.; Hu, S.; Wang, Z.; Gou, J. Effects of ultrasonic vibration in the CMT process on welded joints of Al alloy. *J. Mater. Process. Technol.* **2018**, *259*, 282–291. [CrossRef]
22. Nie, Y.P.; Zhang, P.L.; Xi, G.J.; Hua, Z. Rapid prototyping of 4043 Al-alloy parts by cold metal transfer. *Sci. Technol. Weld. Join.* **2018**, *23*, 527–535. [CrossRef]
23. Kannan, A.R.; Shanmugam, N.S.; Vendan, S.A. Effect of cold metal transfer process parameters on microstructural evolution and mechanical properties of AISI 316L tailor welded blanks. *Int. J. Adv. Manuf. Technol.* **2019**, *103*, 4265–4282. [CrossRef]
24. Su, C.; Chen, X.; Konovalov, S.; Arvind, R.; Huang, L. Effect of deposition strategies on the microstructure and tensile properties of wire arc additive manufactured Al-5Si alloys. *J. Mater. Eng. Perform.* **2021**, *30*, 2136–2146. [CrossRef]
25. Cong, B.; Ding, J.; Williams, S. Effect of arc mode in cold metal transfer process on porosity of additively manufactured Al-6.3%Cu alloy. *Int. J. Adv. Manuf. Technol.* **2015**, *76*, 1593–1606. [CrossRef]
26. Gu, J.; Ding, J.; Williams, W.; Gu, H.; Ma, P.; Zhai, Y. The effect of inter-layer cold working and post-deposition heat treatment on porosity in additively manufactured aluminum alloys. *J. Mater. Process. Technol.* **2016**, *230*, 26–34. [CrossRef]
27. Geng, H.; Li, J.; Xiong, J.; Lin, X.; Zhang, F. Geometric Limitation and Tensile Properties of Wire and Arc Additive Manufacturing 5A06 Aluminum Alloy Parts. *J. Mater. Eng. Perform.* **2017**, *26*, 621–629. [CrossRef]
28. Zhang, C.; Li, Y.; Gao, M.; Zeng, X. Wire arc additive manufacturing of Al-6Mg alloy using variable polarity cold metal transfer arc as power source. *Mater. Sci. Eng.* **2018**, *711*, 415–423. [CrossRef]

29. Horgar, A.; Fostervoll, H.; Nyhus, B.; Ren, X.; Eriksson, M.; Akselsen, O.M. Additive manufacturing using WAAM with AA5183 wire. *J. Mater. Process. Technol.* **2018**, *259*, 68–74. [CrossRef]
30. Gu, J.; Wang, X.; Bai, J.; Ding, J.; Williams, S.; Zhai, Y.; Kun, L. Deformation microstructures and strengthening mechanisms for the wire+arc additively manufactured Al-Mg4.5Mn alloy with inter-layer rolling. *Mater. Sci. Eng. A Struct. Mater. Prop. Misrostructure Process.* **2018**, *712*, 292–301. [CrossRef]
31. Su, C.; Chen, X.; Gao, C.; Wang, Y. Effect of heat input on microstructure and mechanical properties of Al-Mg alloys fabricated by WAAM. *Appl. Surf. Sci.* **2019**, *486*, 431–440. [CrossRef]
32. Ryan, E.M.; Sabin, T.J.; Watts, J.F.; Whiting, M.J. The influence of build parameters and wire batch on porosity of wire and arc additive manufactured aluminium alloy 2319. *J. Mater. Process. Technol.* **2018**, *262*, 577–584. [CrossRef]
33. Ortega, A.G.; Corona, G.L.; Salem, M.; Moussaoui, K.; Segonds, S.R.S.; Deschaux-Beaume, F. Characterisation of 4043 aluminium alloy deposits obtained by wire and arc additive manufacturing using a Cold Metal Transfer process. *Sci. Technol. Weld. Join.* **2019**, *24*, 538–547. [CrossRef]
34. Klein, T.; Schnall, M. Control of macro-/microstructure and mechanical properties of a wire-arc additive manufactured aluminum alloy. *Int. J. Adv. Manuf. Technol.* **2020**, *108*, 235–244. [CrossRef]
35. Li, Z.; Zhu, M.; Dai, T. Simulation of micropore formation in Al-7%Si alloy. *J. Acta Metall. Sin.* **2013**, *49*, 1032–1040. [CrossRef]
36. Yuan, T.; Yu, Z.L.; Chen, S.J.; Xu, M.; Jiang, X.Q. Loss of elemental Mg during wire + arc additive manufacturing of Al-Mg alloy and its effect on mechanical properties. *J. Manuf. Process.* **2020**, *49*, 456–462. [CrossRef]
37. Jiang, J.; Lai, S.; Lu, L.; Fang, J.; Liu, G.; Meng, S.; Jiang, F. Research progress of 5XXX series aluminum magnesium alloy. *Manned Space* **2019**, *25*, 411–418.
38. Yan, X.; Zhao, G. Ac pulse plasma arc welding process of 5052 aluminum alloy. *Welding* **2018**, *02*, 30–34.
39. Huang, B.Y.; Li, C.G.; Shi, L.K.; Qi, G.Z.; Zuo, T.L. *Chinese Materials and Engineering Canon*; Chemical Industry Press: Beijing, China, 2005; p. 88.
40. Ren, L.L.; Gu, H.M.; Wang, W.; Wang, S.; Li, C.D.; Wang, Z.B.; Zhai, Y.C.; Ma, P.H. Effect of Mg content on microstructure and properties of Al-Mg alloy produced by the wire arc additive manufacturing method. *Materials* **2019**, *12*, 4160. [CrossRef]

MDPI
St. Alban-Anlage 66
4052 Basel
Switzerland
www.mdpi.com

Materials Editorial Office
E-mail: materials@mdpi.com
www.mdpi.com/journal/materials

Disclaimer/Publisher's Note: The statements, opinions and data contained in all publications are solely those of the individual author(s) and contributor(s) and not of MDPI and/or the editor(s). MDPI and/or the editor(s) disclaim responsibility for any injury to people or property resulting from any ideas, methods, instructions or products referred to in the content.

www.ingramcontent.com/pod-product-compliance
Lightning Source LLC
LaVergne TN
LVHW070644100526
838202LV00013B/874